OUTER PLANETS

Glenn F. Chaple

Greenwood Guides to the Universe
Timothy F. Slater and Lauren V. Jones, Series Editors

GREENWOOD PRESS
An Imprint of ABC-CLIO, LLC

A B C ☰ C L I O

Santa Barbara, California • Denver, Colorado • Oxford, England

Library of Congress Cataloging-in-Publication Data
Chaple, Glenn F.
    Outer planets / Glenn F. Chaple.
        p. cm. — (Greenwood guides to the universe)
    Includes bibliographical references and index.
    ISBN 978-0-313-36570-6 (hard copy : alk. paper) —
ISBN 978-0-313-36571-3 (ebook)
    1. Outer planets. I. Title.
    QB659.C43 2009
    523.4—dc22   2009019682

13 12 11 10 9   1 2 3 4 5

This book is also available on the World Wide Web as an eBook.
Visit www.abc-clio.com for details.

ABC-CLIO, LLC
130 Cremona Drive, P.O. Box 1911
Santa Barbara, California 93116-1911

This book is printed on acid-free paper (∞)

Manufactured in the United States of America

OUTER PLANETS

Greenwood Guides to the Universe
Timothy F. Slater and Lauren V. Jones, Series Editors

Astronomy and Culture
Edith W. Hetherington and Norriss S. Hetherington

The Sun
David Alexander

Inner Planets
Jennifer A. Grier and Andrew S. Rivkin

Outer Planets
Glenn F. Chaple

Asteroids, Comets, and Dwarf Planets
Andrew S. Rivkin

Stars and Galaxies
Lauren V. Jones

Cosmology and the Evolution of the Universe
Martin Ratcliffe

For my grandchildren, Katie and Sam
May your stars always shine

Contents

Series Foreword

Not since the 1960s and the Apollo space program has the subject of astronomy so readily captured our interest and imagination. In just the past few decades, a constellation of space telescopes, including the Hubble Space Telescope, has peered deep into the farthest reaches of the universe and discovered supermassive black holes residing in the centers of galaxies. Giant telescopes on Earth's highest mountaintops have spied planet-like objects larger than Pluto lurking at the very edges of our solar system and have carefully measured the expansion rate of our universe. Meteorites with bacteria-like fossil structures have spurred repeated missions to Mars with the ultimate goal of sending humans to the red planet. Astronomers have recently discovered hundreds more planets beyond our solar system. Such discoveries give us a reason for capturing what we now understand about the cosmos in these volumes, even as we prepare to peer deeper into the universe's secrets.

As a discipline, astronomy covers a range of topics, stretching from the central core of our own planet outward past the Sun and nearby stars to the most distant galaxies of our universe. As such, this set of volumes systematically covers all the major structures and unifying themes of our evolving universe. Each volume consists of a narrative discussion highlighting the most important ideas about major celestial objects and how astronomers have come to understand their nature and evolution. In addition to describing astronomers' most current investigations, many volumes include perspectives on the historical and premodern understandings that have motivated us to pursue deeper knowledge.

The ideas presented in these assembled volumes have been meticulously researched and carefully written by experts to provide readers with the most scientifically accurate information that is currently available. There are some astronomical phenomena that we just do not understand very well, and the authors have tried to distinguish between which theories have wide consensus and which are still as yet unconfirmed. Because astronomy is a rapidly advancing science, it is almost certain that some of the concepts presented in these pages will become obsolete as advances in technology yield previously unknown information. Astronomers share and value a worldview in which our knowledge is subject to change as the scientific enterprise makes new and better observations of our universe. Our understanding of the cosmos evolves over time, just as the universe evolves, and what we learn tomorrow depends on the insightful efforts of dedicated scientists from yesterday and today. We hope that these volumes reflect the deep respect we have for the scholars who have worked, are working, and will work diligently in the public service to uncover the secrets of the universe.

Lauren V. Jones, Ph.D.
Timothy F. Slater, Ph.D.
University of Wyoming
Series Editors

Preface

Astronomy is one of the most exciting of all the sciences. What other scientific discipline can so well combine beauty (Saturn's rings), mystery (the Orion Nebula), and mind-boggling grandeur (the Milky Way Galaxy)? Whether we regard the rugged cratered surface of our next-door neighbor the Moon, or try to contemplate the unimaginable quantities of energy released by a quasar billions of light-years away, the universe is an adventure story filled with wondrous sights, provocative mysteries, and astonishing superlatives.

It's unfortunate that this exciting story often gets lost in a jumble of technical jargon and formulas. *The Outer Planets* is intended for high school and college-age readers who have an interest in astronomy, but prefer a nontechnical approach to the topic. To that end, every effort has been made to present as complete a portrait of the outer planets as possible without having to resort to hard-to-understand concepts and equations.

In Chapter 1, we'll take a historical look at our evolving view of the solar system and these planets. Chapters 2 and 3 will provide insight into the characteristics that define the Jovian (Jupiter-like) planets and the theories of their formation. Chapters 4 and 5 bring us to the colossal planet Jupiter and its amazing moons. Saturn, its fabled rings, and family of satellites will encompass Chapters 6 and 7. The often overlooked "ice giants" Uranus and Neptune receive due attention in Chapters 8 and 9. Chapter 10 brings us to a new and exciting episode in the science of astronomy—the search for and discovery of planets beyond the solar system. Find out about the so-called "hot Jupiters" that forced planetary scientists to reevaluate existing theories of Jovian planet formation. Chapter 11 describes the Grand Tour, the incredible odyssey of *Voyager 2*, the first and (so far) only space probe to defy the odds and explore all four of the outer planets. Finally, in Chapter 12, we'll review what astronomers have learned about the outer planets to date, look at the investigations currently being undertaken, and preview future exploration of these worlds, their rings, and moons.

The text will be augmented by informative sidebars and illustrations, as well as profiles of astronomers past and present who have furthered our knowledge of the outer planets. Words that appear in boldface type are defined in the glossary. Each chapter will conclude with a list of resources for further information and updates.

Appendices at the end of the book include a table of data for the planets; an analysis of the various materials, elements, and compounds that make up the outer planets; a description of the different tools used by astronomers to study the planets and the range of electromagnetic energies (the electromagnetic spectrum) these tools analyze; and a historical timeline of events that relate to the outer planets. Because astronomy is an ever-evolving science with new discoveries coming to light on an almost daily basis, the appendix concludes with a list of useful resources to keep the reader up-to-date. The appendices are preceded by a glossary of key terms and followed by a general bibliography and a detailed subject index.

Acknowledgments

Writing a book about astronomy—in particular, one dealing with the outer planets, their rings, and their moons—can be a daunting task. To begin with, there is a veritable mountain of facts and figures to present, which increases the potential for error on the part of the author. If no errors are present, there still exists the possibility that what is presented as fact today may quickly prove to be erroneous tomorrow. New information with the potential to refine or make obsolete a particular astronomical fact arrives almost daily from observatories and space-science facilities around the world.

Today, as I was listening to my car radio, a newsman on a local radio station announced the discovery of ring arcs around two of Saturn's small moons. Upon arriving home, I confirmed the news report by checking the NASA Web site, and then made the necessary changes to the manuscript.

As important as up-to-date resources are to the making of a book on astronomy, so are the efforts of the individuals responsible for taking it from the original concept to final publication. My own contribution as writer is just one part of the picture. I would like to thank the following for their roles in getting me involved in this project and helping me to prepare the manuscript or to make appropriate and timely updates:

Greenwood Press senior acquisitions editor Kevin Downing for entrusting me with the task of writing this book.

Greenwood Press senior development editor John Wagner for his guidance in helping me fine-tune the manuscript. John's suggestions were always on target.

Artist Jeff Dixon for providing illustrations that added a visual dimension to the text.

Cadmus Communications senior project manager Randy Baldini for his patient assistance in the final draft of the manuscript.

Astronomy magazine senior editors Michael Bakich and Richard Talcott for their encouragement and support during the preparation of this book.

Stephen James O'Meara for factual input on the Saturn ring spokes observed by him with a telescope several years before the *Voyager* missions confirmed their existence.

xiv • ACKNOWLEDGMENTS

James Bryan, of the McDonald Observatory, for further output on O'Meara's observations of the Saturn ring spokes.

Dr. Thomas Arny, professor emeritus of the University of Massachusetts, Amherst, for helpful suggestions as I prepared the manuscript.

Scott Sheppard, of the Carnegie Institution of Washington, for providing timely information about his search for remote moons circling the outer planets.

Finally (and most importantly), I thank my wife, Regina, who patiently endured the long hours I spent at the computer gleaning information from the Internet or working on the manuscript.

Introduction

August 20, 1977: A Titan III-E Centaur rocket thunders off the launch pad at Cape Canaveral, Florida. On board is a 722-kg probe dubbed *Voyager 2*. Its mission is one of the most complex and ambitious ever attempted by NASA—an exploratory journey to the outer planets Jupiter and Saturn. If all goes according to plan, this Christopher Columbus of spacecraft will complete a "Grand Tour" by continuing on to Uranus and Neptune—the first spacecraft from Earth to explore these remote worlds.

In a small New England town 1,800 kilometers (1,120 miles) to the north, a four-year-old girl named Michelle steps outside with her parents to look at the stars. Like *Voyager 2*, she is about to begin an incredible voyage.

In the bleak, cold outer reaches of the solar system, four planets slowly and silently orbit the distant Sun. Jupiter, Saturn, Uranus, and Neptune— the Jovian, or "gas giant" planets—are colossal worlds that easily dwarf our tiny Earth. Each is surrounded by a thick, turbulent atmosphere, encircled by a complex system of rings, and attended by a swarm of fascinating moons and moonlets. The Jovian planets combine the beautiful with the bizarre, enveloping it in a shroud of mystery. Astronomers cannot adequately describe these giants and their moons without resorting to superlatives. You can only shake your head in silent disbelief upon learning the following:

- Jupiter is so huge that, if hollowed out, it could swallow 1,300 Earths.
- Saturn is as heavy as 95 Earths, yet capable of floating in a vast ocean of water.
- The axis of Uranus is tilted so much that the planet rolls across the sky as it orbits the Sun.
- Wind speeds in Neptune's atmosphere are a dozen times faster than a hurricane on Earth.
- Saturn's main rings span a distance equal to two-thirds the distance from the Earth to the Moon, yet are no thicker than the length of a soccer field.
- Jupiter's moon Io is so volcanically active that it literally turns inside out.

Figure I.1 *Voyager 2* launch. NASA.

- Saturn's planet-sized moon Titan has a dense atmosphere that supports a water cycle like Earth's—only with ammonia.
- Rhea, another of Saturn's moons, has its own ring system.
- Miranda, Uranus's fractured moon, has cliffs 20 kilometers (12 miles) high.

Almost as fascinating as these facts about the outer planets are the stories of the men and women whose dedicated work has helped us learn about them. Consider the following:

- The mysterious spokes in Saturn's rings were observed by an amateur astronomer several years before they were photographed up close by the *Voyager* probes.
- Uranus was discovered by a musician for whom astronomy was a hobby.
- Two mathematicians used pencil and paper to predict the existence of Neptune before it was identified by astronomers using telescopes.
- The mirrors on the giant Keck Telescope in Hawaii, used by astronomers to study weather patterns on Uranus and Neptune, can collect as much light as nearly 2 million human eyes.

Come with us on a Grand Tour of Jupiter, Saturn, Uranus, and Neptune as we follow in the footsteps of the space probe *Voyager 2*, launched by the National Aeronautics and Space Administration (NASA) in 1977. Learn how humans slowly and methodically pieced together an evolving picture of their nature, what we currently know about them, and what we hope to discover in the future. Ultimately, our tour of the outer planets will be carried on by future generations of human explorers who will visit firsthand the outer planets and their moons. Perhaps someday you will be involved in outer planet research. Enjoy the voyage!

1

From a Cave in France to the Jet Propulsion Laboratory: A Historical Look at the Outer Planets

What we know about the four outermost planets of the solar system didn't come about overnight. Our current scientific knowledge concerning Jupiter, Saturn, Uranus, and Neptune is a culmination of thousands of years of visual inspection data collection, and analysis—driven at first by human need and later by intellectual curiosity. From the earliest naked-eye observations by our cave-dwelling ancestors to the latest images beamed by the Cassini Orbiter at Saturn to a team of planetary scientists at the Jet Propulsion Lab in Pasadena, California, we have painstakingly pieced together our present-day picture of the outer planets.

Modern historical records allow us to identify with a fair degree of certainty the names of the astronomers responsible for the discoveries of the remote planets Uranus and Neptune. We even know the dates these planets came to light. Unfortunately, historians will never know the identity of the first person or persons to notice Jupiter and Saturn, nor will they be able to point to the exact date when the discoveries were made. Visible to the unaided eye, both planets were already known by the time the earliest civilizations had developed written languages to document their activities. The genesis of our studies of the outer planets takes us back more than ten thousand years to the last Ice Age.

PREHISTORIC OBSERVATIONS OF THE OUTER PLANETS

The prehistoric inhabitants of a cave near what is now Lascaux, France, cared little about the composition of Saturn's rings or what the weather might be like on Neptune. With the Earth still in the grip of the Ice Age, these hunter-gatherers were concerned almost entirely about their day-to-day survival. In a sense, however, they were among the earliest astronomers.

They almost certainly used the daily and seasonal motions of the Sun to keep track of time. Besides the obvious daytime and seasonal meanderings of the Sun, these early sky gazers would have noticed the recurring changes of the Moon's appearance in the night sky. They may not have had a written language, but the paintings they left on the cave walls speak volumes about their activities. Included with renderings of the animals they hunted, some of which are now extinct, are circular patterns of dots that some modern-day astronomers believe depict the cycles of the Moon's phases. Like the Sun, the Moon allowed prehistoric humans the luxury of keeping track of the passage of time.

Besides the Moon, the Lascaux cave dwellers would have come to recognize distinct patterns of stars in the night sky. Over the years, they would have noticed that these star patterns, forerunners of our modern constellations, appear in the same part of the sky after sunset each evening, but slowly shift their positions with the passing seasons. The seasonal parade of the constellations provided another way for our prehistoric ancestors to keep track of the passage of time.

Our early ancestors came to realize that the Sun and Moon traverse the same star groups (what we now call the **zodiac constellations**). It wouldn't have taken long for them to notice five bright stars that behaved differently from the others. While most stars remain fixed relative to one another, these five were mavericks that appeared to wander among the stars, following the paths of the Sun and Moon across the zodiac. We might agree that whatever individual first noted that two of these wandering stars move slower than the others and sought to assign them identifying names was the discoverer of Jupiter and Saturn.

EARLY CIVILIZATIONS: ASTROLOGY AND ASTRONOMY

With the passing of the Ice Age and a return to a milder climate, humans were able to abandon hunting-gathering and engage in agricultural pursuits and the domestication of livestock. Early civilizations arose and so did a more critical approach to observing the night sky. Calendars based on the apparent motions of the Sun and Moon were developed and refined. Not only were the motions of the wandering stars closely studied, but they were recorded. Written language had become a useful tool for the earliest civilizations. Calendar making and plotting the positions of celestial bodies marked the beginnings of true **astronomy**.

Among the first civilized sky gazers were the Babylonians. Eclipses of the Sun and Moon alarmed them as surely as they frightened their cave-dwelling ancestors. Why did the midday Sun suddenly disappear and turn black? What caused the full Moon to darken and turn a lurid reddish color? Were dragons and demons at work? If only these events could be predicted so that the populace could prepare for them!

Eclipses of the Sun and Moon weren't the only cosmic events the Babylonians noted. They also watched the wandering stars as they progressed in their easterly paths among the stars of the zodiac. Did their positions in various zodiac constellations, as well as those of the Sun and Moon, have significance, perhaps determining the personality of a future king on the date of his birth or controlling events on Earth? What role did the two slowest-moving of the wandering stars play in the cosmic scheme? The Babylonians noticed that there were times when the wandering stars abruptly abandoned their easterly paths and began moving in a westerly direction. Would this backward motion also reverse their influence? The roots of **astrology**, the belief that the positions of the Sun, Moon, and five wanderers might influence human behavior and control earthly events, began to take hold.

With a written language and working calendar, the Babylonians recorded the dates of eclipses and tracked the motions of the planets, which proved to be cyclic in nature. The cause of eclipses wasn't as important as being able to predict when such frightening events might occur and get ready for them. Likewise, the reason how or why the Sun, Moon, and planets might influence earthly events was secondary to being able to know their heavenly positions at future times. Imagine being able to predict future happenings and thus be properly prepared by knowing in advance when Mars would enter a zodiac constellation that would signify an impending war or famine.

You might wonder what astrology has to do with astronomy, especially as it pertains to the scientific study of the outer planets. It was the need by astrologers to know where the Sun, Moon, and planets might be at *future* times that gave astronomy a real purpose. Determining planetary orbits through careful studies of their positions (an exercise in astronomy) would allow for accurate predictions of their future locations, permitting the casting of **horoscopes** (an astrological exercise). For thousands of years, astrology and astronomy would exist side by side, one hardly discernable from the other. Not until the Scientific Revolution of the 1600s would astrology begin to lose favor among scientists.

....

Does Astrology Really Work?

"What's your sign?" In countless boy-meets-girl scenarios of the 1960s and 1970s, this question was a popular opening line. What was the significance of the question? Most of the time, it served as a mere "ice breaker" to initiate a conversation. In many cases, however, it was an attempt to test the astrological compatibility of a potential partner.

A person's "sign," or more precisely "astrological sign," refers to the zodiac constellation in which the Sun was positioned at the moment of his/her birth. If the Sun was stationed among the stars in Gemini on the date of your birth, Gemini became your birth sign. According to astrological belief, you share similar character traits with all people born under the sign of Gemini. The importance of asking someone for his/her sign was the belief that some astrological pairings are more compatible than others. We can only wonder how many potentially ideal relationships never got off the ground because of an unfounded belief that a pair of astrological signs didn't form an ideal match.

In ancient times, astrology was considered to be a real science. By the 1700s, scientific discoveries about the true nature of the universe began to cast serious doubts about the validity of astrology. Today, people in the scientific community pay little attention to astrology, placing it in the same category as fortune cookies and tarot cards.

Yet astrology is accepted by a large percentage of the general population, with daily horoscopes appearing in hundreds of newspapers. Nearly one-third of Americans interviewed in a 2003 Harris Poll believed in astrology. A humorous side note to that survey were data indicating that a belief in astrology fell along political party lines; 19 percent of Republicans who took part in the survey believed in astrology, compared to 40 percent of the Democrats!

Despite the popularity of astrology, there is absolutely no scientific evidence that it works. In fact, there are many good reasons to discredit astrology. If a person born under one of the 12 astrological signs is said to possess leadership traits, why are the birth signs of great political and military leaders evenly scattered among all 12 signs? Similar results occur when we look at the birth signs of famous artists, musicians, and even serial killers. Another argument against astrology is that an individual's horoscope is based on the positions of the Sun, Moon, and planets at the moment of birth. Why doesn't astrology consider the moment of conception nine months earlier, when he/she really began life?

Here is another thought to ponder—if you were born during the early half of December (dates assigned to Scorpius), the Sun wasn't in the Scorpion. In fact, the Sun wasn't in any of the 12 zodiac constellations. It was drifting through the constellation Ophiuchus, the Serpent-Bearer. Imagine being asked what your birth sign is and, instead of proudly proclaiming yourself to be a scorpion, having to admit that you're a snake-holder!

Finally, lest we forget that this book is about the outer planets, we might bring Uranus and Neptune into the picture. Their influences are duly noted by modern-day astrologers, yet the practice of astrology thrived for thousands of years before these planets were discovered. Neptune's gravitational influence on Uranus was detected by astronomers decades before the planet was actually discovered. Astrologers cast horoscopes for thousands of years with no hint or clue that there were more than the five planets they charted.

....

THE GREEKS AND EARLY MODELS OF THE UNIVERSE

The ancient Greeks, like the civilizations that preceded them, recognized the five *planetes asteres* (wandering stars). To them the planets represented nothing less than gods or goddesses. One of them always seemed to be close to the Sun, moving rapidly in and out of the evening (and morning) sky. This speedster was Hermes, swift messenger of the gods. A brilliant star radiating with the loveliest white hue was Aphrodite, goddess of beauty. One of the wanderers shone in a baleful reddish hue—a red drop of blood in the night sky. It represented Ares, the Greek god of war. Two others that moved more slowly in the night sky became Zeus, king of the gods, and Cronus, father of Zeus. When the Romans conquered the Greeks, they assimilated much of Greek culture. Hermes, Aphrodite, Ares, Zeus, and Cronus assumed the Romanized names for the planets that we recognize today—Mercury, Venus, Mars, Jupiter (Jove), and Saturn.

The Greeks were more than mere storytellers, weaving mythological tales about the gods and their adventures. They studied the night sky and tried to make reasonable models of the cosmos, based on what they saw. It was obvious to them that the Earth stood immobile at the center of the cosmos. They reasoned that a charioteer feels a breeze and a sense of movement as he rides along on an otherwise windless day. If the Earth either rotated on an axis or revolved around the Sun, the reasoning went, we would likewise experience a constant wind and sense the motion. The fact that we don't experience a constant wind was logical evidence that the Earth stands still.

Thus, the Greeks formulated a model of the universe that put the Earth in the center, with the Moon, Sun and planets in orbit around it. Because many of the early Greek philosophers believed the heavens to be perfect, all orbits were exactly circular. The planets that moved the slowest must be the furthest away. Jupiter, which takes nearly 12 years to traverse the zodiac, and Saturn with its 30-year cycle were thought to be the most distant of the wanderers. Finally, surrounding the Earth, Moon, Sun, and planets was a hollow, spherical shell containing the fixed stars.

By the first century AD, the Greek astronomer Claudius Ptolemy introduced a complex **geocentric** (Earth-centered) **model** that summarized the Greek idea of the structure of the universe and attempted to explain the occasional backward motions of the planets. Instead of orbiting directly around the Earth, each of the planets was seen to move in a perfectly circular path called an **epicycle** around a fixed point in space called the **deferent**. The deferent, in turn, circled the Earth. Inaccurate though it may be, the **Ptolemaic model** so successfully explained the perceived workings of the solar system that it would stand virtually unchallenged for 14 centuries.

....

Claudius Ptolemy (ca. 85–165 AD)

Claudius Ptolemy was born in the Greek city of Ptolemais Hermii, but spent most of his life in the Egyptian city of Alexandria, then a world center of culture and learning. As is the case with many figures of antiquity, little is known about his life other than his works and astronomical contributions. Ptolemy's chief claim to fame is his publication of the *Megale syntaxis* [Great Collection],

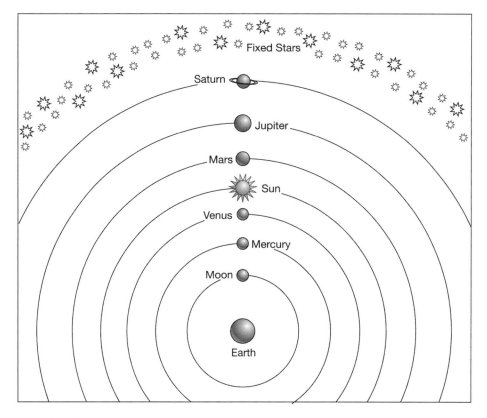

Figure 1.1 Ptolemy's Earth-centered model. Illustration by Jeff Dixon.

better known by its Arabic title, the *Almagest*. It was a 13-book summary of all scientific knowledge of the time.

In the *Almagest*, Ptolemy introduces the geocentric (Earth-centered) model of the solar system to which his name would be attached. The idea that the Earth is the center of the universe was nothing new, but Ptolemy's complex model was. To explain the observed movement of the planets across the night sky, in particular, their occasional backward motion, he devised a system in which the planets orbit in perfectly circular paths around a central point called the deferent, which, in turn, travels in a perfectly circular path around the Earth. Despite the obvious flaws in the Ptolemaic model, it remained largely unchallenged for the next 14 centuries.

THE RENAISSANCE AND A NEW SOLAR SYSTEM

The late sixteenth and early seventeenth centuries were a time of rebirth. Explorers were sailing the oceans and returning to announce the existence of previously unknown lands. Young scholars, their minds fueled by the excitement of change, began to challenge the teachings of their elders. Artists and musicians abandoned traditional methods to adopt new ideas.

Caught up in this time of revolutionary thought was the Polish mathematician Nicolaus Copernicus.

Copernicus questioned the validity of Ptolemy's Earth-centered model, seeing it as far too complex and cumbersome. Through diligent observation of the planets, he formulated a **heliocentric model** that placed the Sun

••

The Mysterious Retrograde Loop

Most of the time, a planet will appear to travel in an easterly path through the background constellations, a movement called **direct motion**. Now and then, however, it will slowly come to a halt and begin moving backward in a westerly direction (**retrograde motion**). The phenomenon, especially noticeable in Mars, Jupiter, and Saturn, baffled early sky gazers.

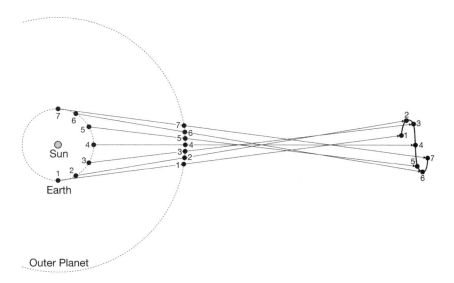

Figure 1.2 Retrograde motion. Illustration by Jeff Dixon.

Their confusion is understandable. Because ancient astronomers believed that the Earth stood still at the center of the universe, they reasoned that any backward motion of a planet must be real. As late as the sixteenth century, most scholars believed that the planets traveled in a way described by the Greek astronomer Claudius Ptolemy 14 centuries earlier—in a complicated system of circular orbits called epicycles and deferents that caused them to appear to move alternately forward and backward.

The establishment of a Sun-centered solar system greatly simplified the planets' orbits and made retrograde motion easy to understand. When we see Jupiter or Saturn moving backward against the backdrop of the distant stars (always at a time when Earth has caught up with them in their orbits and is about to pass them), it's purely an illusion. The planet is still moving forward, but so are we, and at a faster rate. If you're in a car traveling in the passing lane of a highway, any slow-moving vehicle you pass will seem to move backward relative to the distant landscape. Once the car fades into the distance, it appears to resume its forward motion. Retrograde motion is also exhibited by the inner planets Mercury and Venus, but for different reasons.

••

at the center of the universe, with the Earth and other planets in orbit around it. In Ptolemy's model, all heavenly bodies circled the centrally located Earth. In the **Copernican model**, only the Moon orbited the Earth. The beauty of the Copernican model is that it explained the occasional backward forays of the planets in a simple manner.

You might think that Copernicus would eagerly publish the details of his Sun-centered solar system. To the contrary, Copernicus kept his ideas private. For one thing, he feared ridicule, so radical was his heliocentric concept. Of far greater concern was his fear of repercussion from the Roman Catholic Church, which held great power and influence throughout much of Europe. The Church staunchly supported the Ptolemaic model and considered the proposal of anything contrary to be heretical, potentially punishable by death.

Copernicus avoided publishing his theory of a Sun-centered universe until near the end of his life. According to legend, a student brought the first copy of *De revolutionis orbium coelestium* [On the Revolutions of the Celestial Spheres] to Copernicus's death bed. The Polish astronomer reached up, touched the book, and then died.

At the beginning of the Renaissance, little was known about Jupiter and Saturn other than the fact that they were much farther from Earth than the Sun, Moon, and the planets Mercury, Venus, and Mars. Whether you looked at the Ptolemaic model or the Copernican model, Jupiter and Saturn were placed on the outer two circular orbits. Their orbital periods had been established at roughly 12 and 30 years. Their true nature remained unknown.

The telescope changed that, and in the process altered forever our ideas about the order of the solar system. Although no one is completely certain who invented the telescope and when, records lean toward the Dutch spectacle-maker Hans Lippershey, who applied for a patent in 1608. News of this remarkable "toy" that made distant objects appear closer reached the Italian scientist Galileo Galilei the following year, and he quickly set out to fashion one of his own. Making a telescope wasn't an easy task. Galileo had to grind the lenses by hand, set them in a tube, and support the entire optical assembly on a steady mount.

Galileo had good reason to construct a telescope. Unlike those who would use it as a mere toy or as an instrument of war to spy on the enemy from a safe distance, Galileo saw the telescope as a key to unlocking the secrets of the heavens. Like fledgling astronomers today, Galileo christened his crudely made telescope with the most enticing nighttime target—the Moon. Church

Nicolaus Copernicus (1473–1543)

Mikolaj Kopernik, better known by the Latinized name Nicholas Copernicus, was born in Torun, Poland, in 1473. Like many prominent people of his day, Copernicus was a jack-of-all-trades. Besides being an astronomer/mathematician, he was an official in the Catholic Church, a physician, a governor, and a military leader. Astronomy, in fact, was almost a part-time avocation.

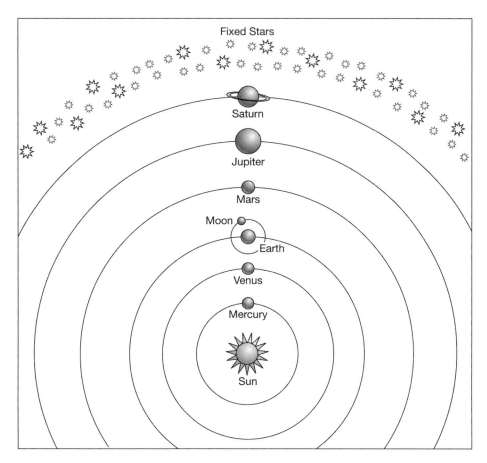

Figure 1.3 Copernicus's Sun-centered model. Illustration by Jeff Dixon.

During his studies at the University of Cracow, Copernicus became convinced that Ptolemy's Earth-centered model of the solar system was incorrect. In 1507, he wrote a short book in which he outlined his belief that the Sun, not Earth, is the body around which all planets orbit. The irony in Copernicus's life is that he was a Church official, yet he agreed with the arguments against the Church-supported geocentric theory. In 1533, he wrote *De revolutionibus orbium coelestium* [On the Revolutions of the Celestial Spheres], which further detailed his Sun-centered theory. The manuscript for *De revolutionibus* was first completed in 1533, but the book did not appear in print until a decade later in the year of Copernicus's death. A rather unassuming man, Copernicus correctly feared the controversy, both scientific and religious, that his theory would create.

doctrine proclaimed that the Moon was flawlessly spherical, and Galileo wanted to see perfection as no human before him had. To his amazement, the Moon proved to be anything but perfect. Its surface was pitted with craters, valleys, and mountains. Clearly, something was amiss!

In January 1610, Galileo turned his telescope on Jupiter. For the first time in history, Jupiter was revealed to be more than just a gleaming stellar speck. In Galileo's telescope, Jupiter presented itself as a distinct disk-like object.

But it was the three bright "stars" that accompanied Jupiter, plus a fourth observed a few nights later, that aroused Galileo's curiosity. Observing them from night to night, he came to the realization that they weren't background stars at all, but bodies physically orbiting around Jupiter. Church dogma stated that all heavenly bodies circle the Earth. The master of these four disobedient little "stars" was Jupiter. Again, something was amiss!

Galileo, a true Renaissance scientist and adherent of a Sun-centered universe, had acquired dramatic visual evidence to support the Copernican model. He published an account of his telescopic findings in a revolutionary book called *Sidereus nuncius* [the Starry Messenger]. Ptolemy's Earth-centered model, which had stood virtually unchallenged for 14 centuries, was at last being put to the test.

The telescope also permitted Galileo the first close-up glimpse of the most distant-known planet, Saturn. He was the first to see Saturn's rings but, because of the inferior optical quality and low magnifying power of his telescope, failed to fathom their true nature. To him, they were "jug handles" firmly attached to Saturn's disk. More detailed observations of Jupiter and Saturn would require improvements in telescope design.

Astronomy changed with the arrival of the telescope. For thousands of years, humans had studied the Sun, Moon, planets, and stars for practical purposes, two examples being to develop calendars to keep track of time and to cast horoscopes for future dates. The telescope ushered in an era of astronomical study fueled by one purpose—curiosity!

As long as Ptolemy's Earth-centered model placed the Sun, Moon, and planets in Earth-centered orbits, astronomers could never attain a true understanding of the solar system. Copernicus had published the idea of a Sun-centered solar system, and Galileo had supplied visual proof, but the nail in the coffin for the Ptolemaic model came not with an astronomer's observation, but with a physicist's definition of one of the most powerful forces in the universe.

In 1687, English physicist Isaac Newton published *Philosophiae naturalis principia mathematica* [Mathematical Principles of Natural Philosophy]. In the *Principia*, Newton outlined his laws of gravity, thereby revealing the existence of the "glue" that keeps the planets in their orbits around the Sun. The Copernican model of the solar system at last began to receive universal acceptance. Jupiter and Saturn, along with Mercury, Venus, Earth, and Mars, were planets in orbit around the Sun.

THE OUTER PLANETS IN THE MODERN ERA OF ASTRONOMY

During the century that followed Galileo's pioneering telescopic forays into the night sky, the design, size, and optical quality of telescopes improved dramatically. Thanks to the visual gifts these instruments bestowed, astronomers were able to make increasingly accurate, detailed observations of

Jupiter and Saturn. The "jug handles" Galileo saw affixed to Saturn's disk proved to be flat, thin rings surrounding the planet. Beyond the rings, a handful of moons were discovered. Storms in the atmospheres of Jupiter and Saturn appeared as dark or light spots, serving as markers that allowed astronomers to gauge the planets' rapid rotation rates. No wonder their disks appeared slightly flattened!

On a late winter night in 1781, the telescope gave astronomy a most unexpected gift—a new planet. On that evening, the unknown English astronomer William Herschel was conducting a telescopic survey of the heavens around the constellation Gemini when he stumbled upon an unusual-looking star. The "star" turned out to be the first planet discovered in modern times—Uranus. Thanks to the telescope, our solar system wasn't just becoming more familiar, it was growing larger.

With the scientific need to attain reasonably accurate calculations of the orbits of the planets, their distances, and sizes, mathematics began to play an increasingly important role as an astronomical tool. Studies of the orbits of their moons betrayed the masses of these planets, and from their masses and volumes, astronomers could calculate their densities. It was obvious that Jupiter, Saturn, and Uranus were far larger and more massive than Earth, but much less dense.

One of the greatest triumphs of mathematics in astronomy was the discovery of Neptune in 1846. Noting that something was amiss with the orbit of Uranus and believing that the gravitational pull of an unseen planet was to blame, mathematicians calculated its position. After a struggle to convince astronomers to point their telescopes in the right location, the new planet, Neptune, was found. Mathematics (and Newton's laws of gravity) helped astronomers discover a new planet.

In the middle of the nineteenth century, two innovations were developed that would revolutionize astronomy—photography and spectroscopy. The photographic plate, which came into use in the mid-1800s, proved valuable in several ways. First of all, it was fast. An astronomer might spend the better part of an hour making a hand drawing of Jupiter's cloud belts as seen through the telescope. A photographic plate could do the job in seconds, and produce far more accurate results. Photography also proved its value by virtue of its greater sensitivity to light too dim for the human eye alone. A number of moons orbiting the outer planets, too faint to be viewed at the eyepiece, were captured on photographic plates.

The second innovation had even more far-reaching effects. In 1835, the French philosopher Auguste Compte stated that astronomers would never learn the composition of the stars and planets. The rationale for this bold statement was simple. Because there was no way to negotiate the vast distances that separate us from cosmic bodies, we would be unable to obtain samples of their material for chemical analysis. Within years of Compte's remarks, astronomers found an indirect way to "taste" the stars and planets—the spectroscope.

The spectroscope works on the principle that light, when passed through a prism, is separated into its component colors. This happens because light travels in waves, and the various colors of light have discreet wavelengths, with red being the longest and violet the shortest. When a beam of pure white light passes through a triangular piece of glass (a prism), the longer wavelengths (red) are bent less than the shorter ones (violet). The light is spread out into a rainbow, or what scientists call a continuous **spectrum**.

During experimentation, scientists discovered that sunlight passed through a prism didn't produce a continuous spectrum. Instead, the spectrum was interrupted by dark lines at irregular intervals. More analyses revealed that the dark lines were produced by elements in the Sun absorbing those particular wavelengths of light. Lab analyses of the spectra produced by various elements turned up an exciting fact. Each element has a unique spectrum. Astronomers might not be able to capture a sample of a distant star or planet, but they could obtain the fingerprints of its elements.

Not only were astronomers able to analyze the chemistry of the stars, but they could sample the atmospheres of the planets from Earthbound observatories. Jupiter and Saturn were found to contain methane and ammonia. Saturn's moon, Titan, was found to have an atmosphere rich in methane.

The spectroscope told us more than just the chemical composition of heavenly bodies. It also revealed their motion. If you've ever stood outside as an ambulance passed, you probably noticed that the pitch of its siren, high as it approached, suddenly dropped as it sped away. What you experienced was the so-called **Doppler effect**. As the ambulance moved toward you, the sound waves coming from the siren were "squeezed together." This effectively reduced their wavelength, creating a high pitch. As the ambulance receded, the sound waves were stretched out, creating a longer wavelength and lower pitch.

Light also travels in waves. Unlike sound waves, which need to be carried by a medium like air or water, light waves can traverse the vacuum of space. Like the sound from a speeding ambulance, light from a star or planet moving toward or away from the Earth will undergo a Doppler shift. The light waves from an approaching celestial object will be squeezed together and shortened, shifting the lines on its spectrum toward the blue (a blue shift). Conversely, a red shift of its spectrum would indicate motion away from us. The faster a cosmic body moves, the greater the blue or red shift of its spectrum. Astronomers were actually able to determine the rotation speed of the outer planets by keying in on the spectrum of the side of the planet turning toward us. Its blue shift indicated the planet's rotation speed.

From prehistoric times through the middle of the twentieth century, astronomers relied solely on visible light to glean the secrets of the cosmos. Studying the heavens only in visible light is as counterproductive as trying to learn about world history by merely looking at events pertaining to ancient Greece. For a complete picture of world history, you need to look at other cultures across a wide spectrum of time. For a complete picture of the

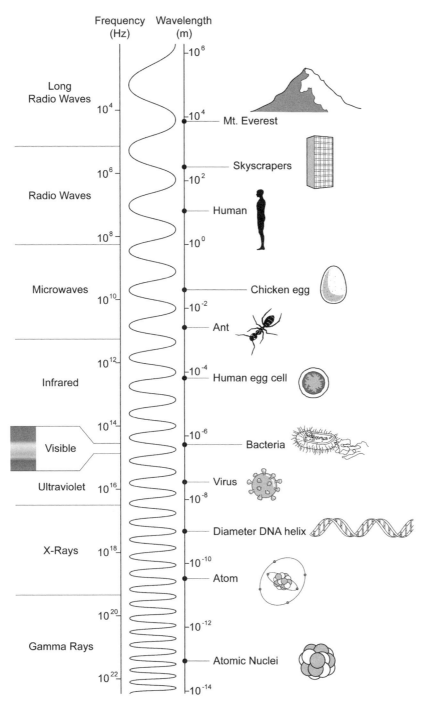

Figure 1.4 The electromagnetic spectrum. Illustration by Jeff Dixon.

universe, you need to look the heavens across a wide spectrum of **electromagnetic energy** (E-M energy).

Visible light represents a tiny fraction of the energies released by stars and planets. Like visible light, these various forms of electromagnetic energy travel in waves. The only difference in the various types of E-M energy, which includes radio energy, microwave, infrared, visible light, ultraviolet, x-rays, and gamma rays, is in their wavelengths. A summary of the different forms appears in a representation of the **electromagnetic spectrum** in Figure 1.4.

Although several of these forms of electromagnetic energy were discovered earlier, the gathering of information by astronomers using wavelengths other than light didn't begin in earnest until the middle of the twentieth century. One of the most important of these nonvisual tools was the **radio telescope**.

When the radio came into general use in the early decades of the twentieth century, annoying background noise was found to infiltrate some frequencies. Bell Laboratories assigned Karl Jansky the task of identifying the source of this static. Jansky discovered that the source of the radio emissions wasn't in our atmosphere, as was believed, but from *space*! At the end of World War II, astronomers began use the first radio telescopes to capture and analyze radio waves from space, to map the sky for important radio sources. The radio telescope was an offshoot of the bowl-shaped radar detectors used during the war. It's an oversized version of the TV satellite dishes that grace the backyards of many homes.

The radio telescope works in two modes. It can *passively* collect radio energy from outer space (radio astronomy), or it can *actively* beam radio signals off nearby solar system targets and analyze the signal that bounces back (radar astronomy). Radio astronomy led to the discovery of Jupiter's magnetic field and powerful storms in the planet's atmosphere. Radar astronomy provided the first clues about the size of particles in Saturn's rings.

Astronomers also began to look at the forms of electromagnetic radiation that bracket visible light—infrared (IR) and ultraviolet (UV). Unlike visible light and radio energy, these two forms of radiation are largely absorbed by the atmosphere. Early IR and UV studies were conducted by placing the detectors on high-altitude balloons.

THE SPACE AGE

During the latter half of the twentieth century, astronomers were able to study the outer planets with several "eyes" that could see in a variety of wavelengths. Still, the view was often blurry. Ground-based instruments were at the mercy of the ocean or air that surrounds our planet. Air turbulence can severely distort the image an astronomer sees at the eyepiece. This is why observatories are routinely constructed on mountaintops, where the atmosphere is thinner and less apt to "play" with incoming starlight. Studies of infrared, ultraviolet, and x-ray energies were also hampered by the atmosphere, which absorbs those wavelengths.

With the launch of the world's first satellite, *Sputnik I*, by the Soviet Union in the autumn of 1957, the Space Age officially began. With it came the opportunity for astronomers to take their equipment where it hadn't gone before—into space.

Many of the earliest satellites were launched for military and communication purposes. By the 1960s, however, scientific satellites were launched that could analyze the infrared and ultraviolet energies being released by bodies in outer space.

Imagine that an astronomer has access to the largest earth-based optical and radio telescopes and a host of Earth-orbiting satellites. Would he or she want more? The answer is an emphatic yes! While all of these instruments will collect useful data, they cannot provide a close-up intimate portrait of a distant planet. To see intricate detail in the atmosphere of Jupiter, map the surface of Saturn's moon Titan, or gauge the magnetic fields of Uranus and Neptune, space scientists needed to "get there" with an interplanetary space probe. In effect, a space probe is a well-equipped observatory sent out to a planet to collect data and relay it back to Earth. Some of the notable interplanetary probes we'll look at in future chapters include *Pioneer 10* and *11*, *Voyagers 1* and *2*, the *Galileo Orbiter* to Jupiter, and the *Cassini-Huygens Orbiter* to Saturn.

Space probes offer tremendous returns, but their high cost and relatively short lifetimes are drawbacks. What if a large telescope could be placed in space high above our atmosphere? This was the idea that led to the development of the Hubble Space Telescope (HST). The concept of an orbiting space telescope had its roots in the 1920s, but 70 years had to pass before technology allowed the HST to operate. The Hubble has been a boon to planetary astronomers. In 1994, it provided high-resolution images of the collision of fragments of Comet Shoemaker/Levy 9 with Jupiter. In recent years, it has looked at evolving weather patterns on Uranus and Neptune.

Despite the advantages of conducting astronomy with space telescopes, satellites, and interplanetary probes, ground-based astronomy is far from dead. Thanks to modern computer technology, two recent advances— **adaptive optics** and digital **charge-coupled device (CCD)** photography, astronomers can image the outer planets in detail that wasn't possible in precomputer times.

Traditional large ground-based telescopes like the 5-meter (200-inch) reflector at Mount Palomar in California were always at the mercy of air turbulence. Ripples in the atmosphere would cause the image to weave and flutter. Modern mega telescopes like the 10-meter (400-inch) Keck Telescope in Hawaii have thin, flexible mirrors that are computer controlled to flex and bend in response to air turbulence, negating its effects.

Just 20 years ago, photographers used cameras loaded with photographic film. Though able to accumulate light much better than the human eye, standard film was still painfully slow, and it had to be taken to a photo shop to be developed (a process that usually required outside help and hours or days of waiting time). To take more pictures, the camera had to be reloaded with new film. Digital (CCD) technology was a boon to astronomers.

If you enjoy taking photographs, you probably have a digital camera. A light-sensitive computerized chip quickly collects the images, which can be placed on a computer screen for immediate viewing. Once the images are downloaded, the chip is free to be used over and over.

Digital photography has become a major tool of present-day astronomy, replacing the old-fashioned photographic plate the way digital cameras have supplanted film-loaded cameras. CCD astronomy has become popular with amateur astronomers, as well as with the professionals. Using CCD technology, amateur astronomers are taking high-resolution images of the planets that rival the best photographic plates made by professional astronomers a few decades ago.

By using a CCD camera with one of the large adaptive optics telescopes, astronomers can image the outer planets in detail never before imagined. CCD imaging made possible numerous recent discoveries of faint moons orbiting the outer planets, and it's been a key ingredient in monitoring the evolving climates of Uranus and Neptune.

There is more information on the various tools astronomers use to study the outer planets in Appendix C in the back of the book. A historical time-line summarizing the major discoveries that relate to the outer planets appears in Appendix D.

RECOMMENDED READING

Pannekoek, A. *A History of Astronomy.* New York: Interscience Publishers, Inc., 1961.

A detailed look at the history of astronomy.

Sagan, Carl. *Cosmos.* New York: Random House, 1980.

A highly entertaining book on astronomy. Includes numerous historical excerpts.

WEB SITE

www.badastronomy.com/bad/misc/astrology.html.

Phil Plait's Bad Astronomy Web site takes a critical look at astrology and provides links to similar Web sites.

2

The Nature of the Beast: What Is a Jovian Planet?

SUPERIOR PLANETS: IT'S ALL ABOUT THE ORBIT

We can define our Sun's family of planets two ways—by their orbital positions relative to Earth and by their physical characteristics. The former method of grouping was accomplished centuries ago, once the Sun was established as the center of the solar system and the relative order of the planets' orbits was determined. Mercury and Venus, whose orbits lie closer to the Sun than Earth's, are called **inferior planets**. Those whose Sun-circling paths are farther out (Mars, Jupiter, Saturn, Uranus, and Neptune) are collectively known as the **superior planets**.

To better understand how the orbits of inferior and superior planets determine their visibility in our sky, imagine for a moment that the Earth is perfectly still in space and that the daytime sky is dark enough to allow us to see the planets. Gazing sunward from our earthly platform, we see the inferior planets Mercury and Venus orbiting the Sun. Because the orbits of all the planets lie in approximately the same plane, we get an edge-on view of their orbital antics. Mercury, being closer to the Sun, moves faster and never strays more than 27° to either side. Slower-moving Venus can wander up to 47° from the Sun—well out of the reach of its overwhelming glare.

If we could travel out in space far above Earth's north pole and look downward, our panoramic view of the solar system would show Mercury and Venus (and all of the planets) circling the Sun in a counterclockwise direction. Back on Earth, we notice that Mercury and Venus seem to weave

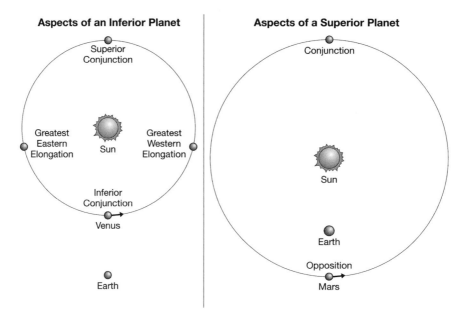

Figure 2.1 Aspects of orbits of inferior and superior planets. Illustration by Jeff Dixon.

back and forth from one side of the Sun to the other, alternately passing in front of and behind our daytime star. They travel from right to left (west to east) when on the Sun's far side, and from left to right (east to west) when passing between us and the Sun.

Although Mercury and Venus are closest to Earth when they pass between us and the Sun (a position called **inferior conjunction**), this is obviously not a good time to view them. **Superior conjunction** on the Sun's far side isn't any better. During conjunctions, Venus and Mercury are hidden by the Sun's glare. The best times to view the inferior planets are when their orbits carry them as far from the Sun as possible (**greatest eastern and western elongations**). At the time of greatest eastern elongation, Mercury and Venus set after the Sun, and are briefly visible above the western horizon as evening planets. Conversely, a western elongation places the planet ahead of the Sun, bringing it into view above the eastern horizon before sunrise.

The situation is simpler with the superior planets. Since all of them show the same aspects, we'll concentrate on Jupiter. We'll start with the Earth and Jupiter on opposite sides of the Sun. This configuration is similar to a superior conjunction of Mercury or Venus. However, since none of the outer planets can ever pass between us and the Sun and have an inferior conjunction, we simply call Jupiter's disappearance behind the Sun a **conjunction**.

As with the other planets, Earth and Jupiter orbit the Sun in a counterclockwise direction. Because Earth moves faster (it's on the inside track), Jupiter emerges from the solar glare on the western side of the Sun. It would be seen as a morning planet, rising ever so slightly ahead of the Sun. Day after day, as Earth catches up with the slower-moving Jupiter, we see the

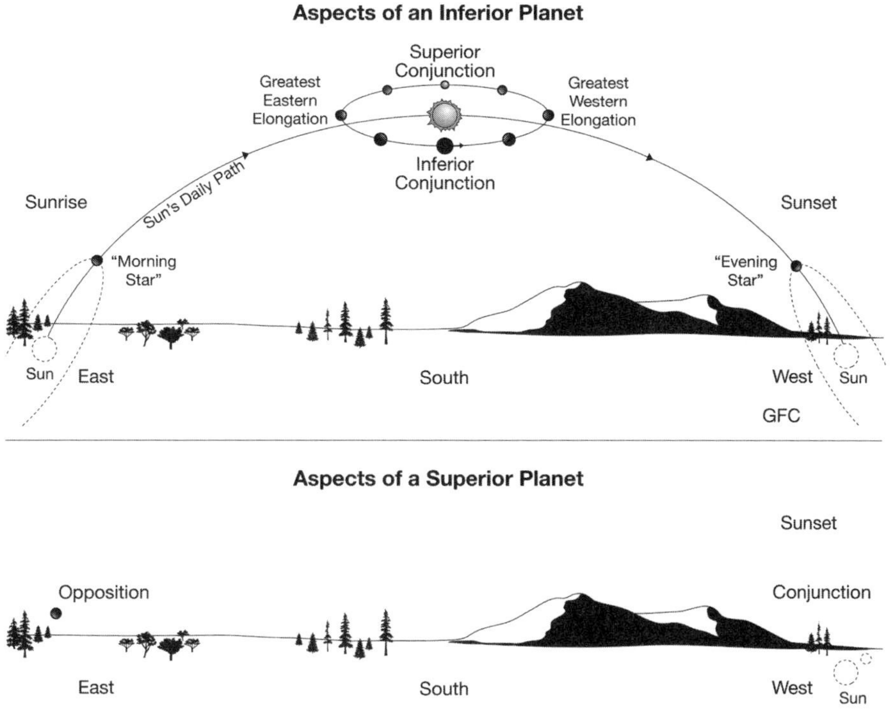

Aspects of an Inferior Planet

Aspects of a Superior Planet

Figure 2.2 Aspects of orbits of inferior and superior planets. Illustration by Jeff Dixon.

Jupiter-Sun gap increase. Eventually, Jupiter is on the opposite side of the sky from the Sun, setting as the Sun rises, and then reappearing at sunset. At this point, called an **opposition**, the Earth has caught up with Jupiter and the two are lined up with the Sun. An opposition is the best time to view a superior planet, because the planet will be visible for most of the night. An opposition also places us as close as possible to a superior planet. After the opposition, the Earth begins to pull away from the superior planet. Each night, the planet will appear closer and closer to the setting Sun until the next conjunction.

Which Way Is East?

When you look at a map, east is to your right, west to your left, and south is at the bottom. Yet when you stand outside facing south and look up at the night sky, east and west are reversed. What's going on?

It's all in your perspective. When you look at a map, it's as though you were in outer space, gazing downward at the landscape. If you were to turn around and lean your back against the map, it would be the same as lying on the ground outside with your feet facing south. Now directions on the map would be reversed, with places to the east on your left.

So, on a map, anything east of the United States is to your right. Outside, a planet shining above the east horizon would be visible to your left.

PHYSICAL CHARACTERISTICS: TERRESTRIAL VERSUS JOVIAN PLANETS

Besides sorting the planets by their orbital positions relative to Earth, we can group them more scientifically by comparing their physical characteristics. This method of planetary classification wasn't possible until the last two centuries, when the telescope and spectroscope began to reveal basic information about their size and composition.

While each of the solar system's planets is unique, astronomers, backed by physical data, can sort them into two distinct groups. The first is composed of the so-called **terrestrial planets**, typified by Earth. The remaining terrestrial planets are Mercury, Venus, and Mars. These four differ greatly from the **Jovian planets**, whose chief representative is Jupiter. Jupiter is joined by Saturn, Uranus, and Neptune. These Jovian planets are, of course, the topic of this book. You'll want to refer to our companion book *The Inner Planets* for more in-depth information about the terrestrial planets.

A glance at the orbital diagram of the planets shown in Figure 2.3 will reveal the first distinction between the terrestrial and Jovian planets. The former are bunched together near the Sun, while the Jovians orbit much farther out. A planet's proximity to the Sun results in several characteristic properties.

First of all, a planet near the Sun receives a greater quantity of its heat and light. Temperatures at the surfaces of the terrestrial planets are far warmer than at the cloud tops of the Jovian planets.

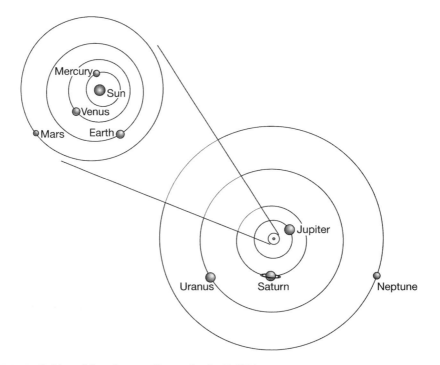

Figure 2.3 Orbits of the planets. Illustration by Jeff Dixon.

The nearer a body is to the Sun, the more it's at the mercy of the Sun's gravitational pull. The terrestrial planets have to move more swiftly in their orbits to avoid being pulled into the Sun. Mercury, nearest of the eight planets, races around the Sun at an orbital speed of 48 kilometers (30 miles) per second, nine times faster than Neptune, the most remote (and slowest-moving) planet.

A fast orbital speed and smaller orbital track combine to yield a short **period of revolution** around the Sun (a planet's "year") for a terrestrial planet. Mercury completes a full orbital cycle in a mere 88 Earth days. At the other extreme, no human could hope to live for one Neptunian year. A jaunt around the Sun for this planet requires 60,190 days (165 Earth years).

The scale rendering of the planets in Figure 2.4 shows another obvious difference between terrestrial and Jovian planets—their size. The Jovian planets are sometimes referred to as the "**gas giants**" (or, as we'll see later, as "**ice giants**") for good reason. Jupiter, Saturn, Uranus, and Neptune boast truly colossal dimensions. Neptune, the smallest Jovian planet, is still four times larger in diameter than Earth, the largest terrestrial planet. Being substantially bulkier, the Jovian planets also have more **mass**. Uranus, the least massive of the "Big Four," carries the mass of over 14 Earths. Giant Jupiter has a mass greater than 318 Earths piled together.

A small planet like Earth might be expected to spin, or rotate, much more rapidly on its axis than a large, more massive one. Oddly enough, the **period of rotation** on a terrestrial planet (its "day") is much longer than one on Jupiter or the other Jovian planets. Jupiter rotates once on its axis in a mere 9 hours, 56 minutes. During one 24-hour Earth day, Jupiter experiences nearly 2.5 days. The terrestrial planet with the slowest rotation is Venus. There, a "day" is eight Earth months long.

Small in size, with somewhat weak gravity fields, and bombarded by the Sun's radiant energy, the terrestrial planets are surrounded by sparse atmospheres, if any at all. The atmospheres of Venus, Earth, and Mars (Mercury is essentially airless) are dominated by heavy gases like carbon dioxide,

Figure 2.4 Relative size of planets. Illustration by Jeff Dixon.

nitrogen, and oxygen. The Jovian planets, by contrast, are enveloped by deep atmospheres rich in the light gases hydrogen and helium, hence their nickname the "gas giants."

The term *gas giant* is misleading, however, because the Jovian planets are in a gaseous state for only a few hundred kilometers below the upper cloud tops. Deeper down, increasing pressure compresses and liquefies the gases to form a slushy ocean. This leads to another fundamental difference between terrestrial and Jovian planets. Unlike the terrestrial planets whose atmospheres rest atop a solid planetary exterior, the Jovian planets have no definite surface. This spells bad news for future astronauts. Jupiter, Saturn, Uranus, and Neptune have no solid terrain on which a spacecraft could land. In recent years, astronomers have cited the presence of water, ammonia, and methane "ices" in Uranus and Neptune as reason to reclassify these "gas giants" as "ice giants."

The Jovian planets are many times more massive than their terrestrial counterparts, but not as much as their huge volumes might indicate. They are composed almost entirely of light materials like hydrogen and helium. Lacking the high concentrations of silicates (rock) and metal found in the terrestrial planets, Jupiter and its brethren gas giants are huge masses of cosmic fluff! As a result, the Jovian planets don't have the **density** of the terrestrial planets. The terrestrials have mean densities 3 to 5.5 times greater than water. Jupiter, Uranus, and Neptune are only slightly denser than water. Saturn, with a mean density just 0.7 that of water, would actually float on a gigantic ocean.

The terrestrial and Jovian planets also differ in what orbits around them. While the four terrestrial planets claim a combined total of three moons, or **satellites** (one for Earth, two for Mars, and Mercury and Venus are moonless), each of the Jovians is attended by a veritable swarm of moons (63 for Jupiter alone). In addition to their impressive families of moons, Jupiter, Saturn, Uranus, and Neptune are each encircled by rings of cosmic debris. For a graphic summary of the characteristics of terrestrial and Jovian planets, refer to the data table for the solar system's eight planets, which appears in Appendix A.

What Happened to Pluto?

"Mary's Violet Eyes Make John Stay Up Nights—Period." How many school children used this mnemonic to learn the names of the nine planets in their proper order: Mercury, Venus, Earth, Mars, Jupiter, Saturn, Uranus, Neptune, and Pluto? There was a 20-year period of confusion beginning in 1979, when Pluto's highly eccentric orbit took it inside Neptune's and the two swapped places in the planetary order. Once Pluto returned to its normal, most-distant position in 1999 (where it would remain for the next two centuries) all was well. Or was it?

When Pluto was discovered on a photographic plate taken by astronomer Clyde Tombaugh in 1931, it was so remote that astronomers were able to ascertain little about its nature other than its orbit. Little by little, however, astronomers began to put together a picture of a world only half the size of our Moon. Its orbit, more eccentric and inclined than those of the other planets, didn't fit the planetary mold. Pluto was found to have three moons of its own, but many asteroids were also found to have satellites, and they hardly qualify as planets.

In the latter part of the twentieth century, astronomers using highly sensitive photographic means began detecting bodies beyond Pluto's orbit that were close to Pluto's size. These objects became known as "trans-Neptunian" or "Kuiper Belt" objects. In 2003, astronomers detected a trans-Neptunian object that proved to be larger than Pluto. In the summer of 2006, astronomers at a meeting of the International Astronomical Union (IAU) drew up a set of definitions for what constitutes a planet. Pluto failed to make the cut. Today, Pluto is officially classified as a "dwarf planet."

In the minds of many who grew up with a nine-planet solar system, the issue is far from resolved. You might even encounter a bumper sticker that reads, "HONK IF PLUTO IS A PLANET."

What Are Planets Made of?

We've all heard the nursery rhyme that proclaims boys are made of "rats and snails and puppy dog tails," while girls are made of "sugar and spice and everything nice." A perusal of any biology text will quickly put that idea to rest! What about the planets? What are they made of?

In the simplest terms, we say the planets are comprised of three basic ingredients: rocky silicates and metals, gases, and ices (like water, methane, or ammonia). These terms can sometimes be misleading when we consider their normal physical states (solid, liquid, or gas) here on the Earth's surface. In the frigid far reaches of the solar system, all of these materials would appear as solids. Their physical states would change in the depths of a planet's interior, where increasing temperature and pressure would first liquefy, and then solidify, them. Here on Earth, silicates and metals will be solid on the surface. Plunge into the Earth's mantle or deeper yet into the outer core, and high temperatures liquefy them. In the still-hotter core, metals become solid once again because of the great pressure found there.

So it is with the other planets, including the Jovians. Ices and gases only exist in solid or gaseous state in the upper levels of their atmospheres. Farther down, both are compressed into liquid state.

Why is it that our Sun's family of planets exists in two distinct zones, with small, dense rocky planets near the Sun and huge, gaseous planets farther out? The answer to this question is found in the processes that created the solar system.

WEB SITE

http://www.nineplanets.org/overview.html.
Bill Arnett's highly useful Nine (8) Planets Web site provides a graphic comparison of terrestrial and Jovian planets.

3

In the Beginning: The Birth of the Jovian Planets

SOLAR SYSTEM GENESIS: EARLY IDEAS

Five billion years ago, there was no solar system. How did our Sun and its attendant planets, moons, asteroids, and comets come to be? Why are those inner denizens, the terrestrial planets, small, dense, and rocky, while the Jovians farther out are much larger and comprised of lighter substances? Astronomers can only speculate, their hypotheses based on studies of the nature and composition of the Sun and other solar system bodies and by the observation of star and planet formation elsewhere in our galaxy.

Science-based theories of the genesis of the solar system are relatively recent in origin. That's because no sensible hypothesis for the formation of the Sun and its family of planets could exist as long as prevailing thought placed the Earth at the center of the solar system. The appropriate climate for the creation of realistic models of solar system formation didn't exist until well into the 1700s when the Copernican Sun-centered model of the solar system came into general acceptance.

The foundation of our present-day ideas of how the solar system came to be lies with the **nebular hypothesis**, which states that the Sun and planets were spawned in a **nebula**, or cosmic gas cloud. One of the earliest to propose this idea was the Swedish scientist/philosopher/theologian Emanuel Swedenborg, who described the theory in 1734. The German philosopher Immanuel Kant in his *General History of Nature and Theory of the Heavens*, published in 1755, expanded the concept. In France, the brilliant mathematician/

astronomer Pierre Simon LaPlace independently conceived of the idea, publicizing his theory in 1796.

In LaPlace's model, the solar system was born when a nebula began to rotate and contract. As it did, the hot central core coalesced into a **star** (the Sun), while rings of gas broke off one by one and cooled to form the **planets**. According to LaPlace, the planets furthest from the Sun were the first to form; those closest to the Sun were the youngest.

··

Pierre-Simon LaPlace (1749–1827)

Pierre-Simon LaPlace was a French mathematician and astronomer, perhaps best remembered for his contribution to the nebular hypothesis—an attempt to explain how the planets in the solar system formed. Born in Beaumont-en-Auge, Normandy, on March 23, 1749, LaPlace originally studied for a career in the Church. At age 16, he was sent to Caen University to study theology, but quickly developed an interest in mathematics.

Figure 3.1 Pierre-Simon Laplace. Images copyright History of Science Collections, University of Oklahoma Libraries.

His genius for mathematics is related in a story suggesting that the 19-year-old LaPlace was sent to Paris by one of his math teachers to meet the famous mathematician Jean le Rond d'Alembert. As a test, d'Alembert assigned LaPlace a difficult math problem with instructions to return with the answer within a week's time. LaPlace showed up with the solution the next day!

LaPlace was elected to the Academy of Sciences in 1773. From that time until his death on March 5, 1827, he devoted much of his effort to producing papers devoted to celestial mechanics. Among his works was the book *Système du Monde*, published in 1796, in which he detailed his nebular hypothesis.

It's interesting to note that LaPlace once contemplated the idea that a small, extremely dense body could possess so much gravity that even light couldn't escape it. LaPlace had described a "black hole" well over a century before astrophysicists began generating mathematical models of such bizarre objects.

••

In science, so it seems, no hypothesis goes unchallenged. In 1778, the French cosmologist Georges-Louis Leclerc, Compte de Buffon, suggested that the planets formed when comets slammed into the Sun, releasing material that cooled and condensed into bodies orbiting the Sun. The **catastrophic hypothesis** had another adherent in the New Zealand astronomer Alexander Bickerton who, in 1880, replaced the comets with a grazing impact from a passing star. A stream of gases ejected in the process would have formed clumps of matter that cooled to form the planets. Modifications of Bickerton's idea were introduced in 1905 by the American astronomers Thomas Chamberlin and Forest Moulton of the University of Chicago, and a decade later by the British astronomers James Jeans and Harold Jeffreys.

The mechanics of these two early and competing theories of planet formation are dramatically different. The nebular hypothesis proposes that a contracting, spinning gas cloud gives birth to a star and its planetary retinue. This scenario would be a common occurrence throughout the galaxy, accomplished each time a star is born. The cataclysmic hypothesis, on the other hand, requires a near collision between two stars. Such an event, considering the great distances that separate the stars, would be exceedingly rare.

As astronomers contemplated the possibility of life elsewhere in the universe, a situation that mandates the existence of planets, they realized that the odds for the existence of extraterrestrial life are mathematically slim if planets were formed only by stellar encounters. If the cataclysmic hypothesis were true, we would likely be alone in the universe.

Fortunately for E.T. movie fans, the catastrophic hypothesis began to unravel in the mid-twentieth century. American astrophysicist Lyman Spitzer argued that any gas drawn out from the Sun would rapidly disperse into space, rather than condense into planets. The recent discovery of disks of gas and dust surrounding young stars and the discovery of exoplanets in relative abundance (see Chapter 11) has led to the demise of the catastrophic hypothesis and a refinement of the nebular hypothesis.

SOLAR SYSTEM GENESIS: THE MODERN IDEA

Imagine that you can travel back in the past—4.6 billion years, give or take a few hundred million years. You expect to see empty space, but instead find our Milky Way Galaxy much as it appears today. The solar system wasn't created at the same instance as the rest of the universe. At the time our Sun and its planets formed, the universe, including our galaxy, was already inconceivably old, having formed in the Big Bang 9 billion years earlier.

The solar system took millions of years to form, so we'll "fast forward" time. A million years passes each minute. The galaxy begins to move—a stately spinning motion that requires over 200 million years for a complete revolution. Looking around, you see a dark cloud looming before you, obscuring the stars beyond. This is the nebula that will ultimately form the solar system. It's mostly comprised of hydrogen and helium gas, interspersed with traces of other elements. It doesn't look like a very promising stellar birthplace, but something is about to happen. There is a blinding flash of light as a nearby red supergiant star, undergoing its final death throes, suddenly erupts in a colossal supernova.

The supernova has two effects on the dark nebula. Shock waves from the explosion compress its gases, creating relatively dense pockets of matter. Meanwhile, gaseous remnants of the supernova "seed" the nebula with a rich assortment of heavier elements. The gravity fields of these pockets allow them to accrue material from the surrounding nebula. As they gather matter, their gravity increases, accelerating their growth.

One of the dense pockets in the nebula will eventually develop into the Sun and solar system. As matter falls into this pocket, it begins to spin, flattening out into a disk-shaped mass. The more the disk contracts, the faster it rotates, much as a figure skater spins faster and faster as she draws her arms inward. The central part grows larger and hotter to form a **protostar** or, in this instance, protosun. All of this happens within the first minute, but we have to remember that what we are witnessing took about 1 million years.

Around the protosun, particles of dust and gas in the spinning mass begin to accrete, or clump together, like snowflakes into pebble-sized particles that continue to grow as they merge. Radiant energy from the embryonic Sun vaporizes and pushes away the light gases from the inner solar system, leaving behind dense metals and rocky material that will continue to accrete to form a swarm of city-sized **planetesimals**. About 800,000,000 kilometers (500,000,000 miles) out is a "frost line," where lighter "ices" like water, ammonia, and methane can exist as solids. They augment the vast quantities of hydrogen and helium already present here. The resulting planetesimals are comprised of these materials.

At last the temperature and pressure in the core of the protosun reaches a critical point where thermonuclear fusion of hydrogen into helium can occur. With a sudden burst of heat and light, the Sun is born. A powerful stream of radiant energy rushes outward, pushing light gases into the outer reaches of the solar system. Near the Sun, the dense, rocky planetesimals undergo a violent period of collision and merger. After 10 minutes (10 million years), the four terrestrial planets have formed and assumed reasonably circular orbits. Because there wasn't very much rocky or metallic material in the original nebula, these planets are rather small.

This scenario for the formation of the terrestrial planets is known as the **core-accretion** model. It's the one most astronomers ascribe to today. Until the early 1990s, it was the generally accepted theory for the formation of the Jovian planets as well, with the understanding that the material forming them was gaseous and more abundant. Some models indicated that Jupiter's core has a mass equal to three to five Earths—too small to gravitationally accrete the material that would form such a large planet before the newborn sun's radiant energy pushes light planet-building material into distant space.

The discovery in the 1990s of Jupiter mass planets in tight orbits around stars beyond our solar system shed further doubt on the core-accretion model. The **disk instability model**, also known as the **gravity instability model**, was created to explain these discrepancies. This model states that gravity waves quickly cause the gases in the disk encircling a protostar to assemble in a spiral pattern. Gaseous material is able to accrue into Jovian mass planets in a matter of a few thousand years. Heavier elements that form the core are added later by cometary impacts. Gravitational interactions between the newborn Jovian planets would fling some inward into orbits close to the parent star.

The disk instability model seems to explain the existence of the Jupiter mass stars found orbiting closely to nearby stars. The formation of the Jovian planets in our solar system is more open to debate. A recent computer-generated model of Jupiter's interior shows a core mass equal to between 14 and 18 Earths. That's enough mass to produce the high gravity needed to form a planet quickly by core accretion. Further studies of the make-up of the Jovian planets, the orbital characteristics of Jovian mass planets orbiting nearby stars, and the disks surrounding extremely young stars will help decide which model, core accretion, disk instability, or a composite of both, is accurate.

"THE WANDERING PLANETS"

In Chapter 1, we noted that early skygazers referred to the planets as "wandering stars." Studies of the compositions of the outer planets and the orbits of icy bodies in the Kuiper Belt in recent decades lend credence to the idea that they truly did wander during the early stages of solar system formation, altering their orbits, and, possibly, even trading places.

To begin with, the current orbits of Uranus and Neptune exist in an area that would have been too material-poor for them to have formed 4.6 billion years ago. They had to assemble closer to the Sun. Some theoreticians believe that gravitational interactions with Jupiter and Saturn, as well as encounters with icy planetesimals, caused the newly formed Uranus and Neptune to move further from the Sun. A few even assert that Neptune started out closer to the Sun than Uranus and actually plowed past Uranus to its present orbital location. It's possible that Saturn also wandered slightly further from the Sun. Jupiter, slowed by friction from material in the solar nebula, may have drifted inward to its present location.

The Age of the Solar System: How Do We Know?

We base our estimate of the solar system's age by studying meteors that fall to Earth from space. Why not just date the rocks found here on the Earth? The main reason is that because of the dynamic nature of our planet's crust, rocks formed when the Earth first formed have long since been lost. We assume that meteorites formed in the earliest days of the solar system's existence. Radioactive dating of elements in these meteorites gives an age of approximately 4.6 billion years.

You're a Star—Literally!

Here's something to ponder next time you're outside on a clear star-filled evening. Your body, like all substances in the universe, is made up a complex mix of elements and compounds, which are comprised of atoms. The simplest element in your body—indeed, in the entire universe—is hydrogen. Hydrogen is the most common element in the universe and the major component of stars. It, along with much of the universe's store of helium, formed in the wake of the Big Bang.

The only way for an element heavier than hydrogen or helium to form is by the fusing together of atoms under extreme heat and pressure. Such conditions exist in the cores of stars, where hydrogen atoms are fused together to form helium and energy. In its final stages, a star creates even heavier elements in its core. Under the extreme heat and pressure of a supernova explosion, heavy elements like iron are formed and released into space.

The elements in this gaseous supernova remnant eventually encounter and mix with the elements in a nearby nebula. These gases are gravitationally reassembled to form new stars, the planets, moons, and other bodies that orbit them, and the life forms that emerge on any planets that offer favorable conditions. That includes—you!

Think of it. The calcium in your bones, the iron in your blood, and the carbon in the proteins that form your muscles were all created in the core of a dying star billions of years ago. You are literally made of star stuff! It's rather ironic that you, a collection of atoms created in a star that went supernova billions of years ago, can gaze upwards and ask, "Why did I explode?"

WEB SITE

http://stardate.org/resources/ssguide/planet_form.html.
This online site of the McDonald Observatory in Texas is a treasure trove of useful astronomical information.

4

Jupiter, King of the Planets

July 9, 1979: After spending nearly two years in space, *Voyager 2* is gliding 570,000 km (350,000 miles) above the cloud tops of Jupiter. Even though *Voyager 2* is 50 percent farther from Jupiter than the Moon is from Earth, the planet appears as large as 30 full Moons placed side by side. On-board cameras, which have been photographing Jupiter for several months, continue relaying images back to Earth. If *Voyager 2* survives its trek through the intense radiation belt that surrounds Jupiter, it will continue onward to Saturn.

Michelle's adventure continues. Now six years old, she will enter first grade in the fall. She loves exploring the night sky, especially through her father's telescope. Jupiter, attended by four sparkly little stars, is one of her favorites.

Jupiter Data

Period of Revolution	11.9 years
Period of Rotation	$9^h 56^m$
Axis Tilt	$3.1°$
Equatorial Diameter	142,984 km (88,848 mi) = 11.2 Earths
Volume (Earth = 1)	1321
Mass (Earth = 1)	317.8
Surface Gravity* (Earth = 1)	2.7
Density (water = 1.0 g/cm^3)	1.33 g/cm^3
Number of Moons	63
Mean Distance from Sun	5.2 AU** (778,570,000 km [483,780,000 mi])

* Since none of the Jovian planets has a solid surface, gravity is calculated at the visible cloud tops.

** An AU (Astronomical Unit) is the average distance of the Earth from the Sun.

JUPITER: AN EARLY VIEW

Thousands of years ago, a bright, pale-yellow "wandering star" caught the attention of early Greek sky gazers. Perhaps because of its brilliance (only the Sun, Moon, and Venus shine brighter) and the regal way it meandered through the constellations of the zodiac, they named it Zeus in honor of the king of their gods. Modern astronomy has shown that Zeus (now bearing the Roman name Jupiter) is the largest planet in the solar system. In size, Jupiter is truly King of the Planets. Even without the benefit of the telescope, the ancient Greeks had it right all along.

The Greeks may have given the planet an appropriate name, but they didn't discover it. One of the five "wandering stars," Jupiter has been known since antiquity. It would be hard for even the most unobservant person to miss seeing Jupiter in the night sky, especially on those occasions when the Moon happens to be passing by it and the two form an eye-catching pair. Jupiter typically shines at **magnitude** −2.5, three times greater than any of the nighttime stars and bright enough to be faintly visible in the daytime sky. It's quite reasonable to assume that knowledge of Jupiter's existence was handed on to the earliest civilizations by their prehistoric forebears.

Figure 4.1 Jupiter. AP Photo/NASA Jet Propulsion Laboratory/University of Arizona.

Little of Jupiter's true nature can be ascertained with the unaided eye. Other than plotting its changing position in the nighttime sky, determining Jupiter's 12-year cycle through the 12 constellations of the zodiac, and coming to the logical conclusion that only Saturn was more remote among the five wandering stars, early sky gazers knew very little about Jupiter.

....

How Bright? The Magnitude System

When astronomers want to describe how bright an object shines in the night sky, they refer to its magnitude. The magnitude system used today dates back more than 2,000 years to the Greek astronomer Hipparchus. He identified the 20 brightest stars in the sky as being "first magnitude" in brightness. Stars slightly fainter were designated as "second magnitude." Hipparchus continued the sequence down to "sixth magnitude," the faintest stars visible on a clear, moonless night.

In 1856, the English astronomer Norman Pogson revised the Hipparchus magnitude scheme to provide the greater accuracy modern astronomy demanded. In Pogson's updated system, a difference of five magnitudes equals a brightness ratio of precisely 100:1. In other words, a magnitude 1.0 star would shine a hundred times brighter than a star of magnitude 6.0. A difference of a single magnitude would represent a brightness ratio of 2.512. A star of magnitude 1.0 is approximately 2.5 times brighter than a 2.0 magnitude star, and $2.5 \times 2.5 = 6.25$ times brighter than a star of magnitude 3.0.

Because the telescope allows us to see stars fainter than the naked eye limit, we encounter brightnesses far lower than Hipparchus's sixth magnitude. Standard binoculars can reach 9th magnitude, while stars as faint as 12th or 13th magnitude can be glimpsed in small backyard telescopes. The sharp "eye" of the Hubble Space Telescope has captured images of galaxies in deep space shining at 30th magnitude—fainter than a candle viewed from the distance of the Moon.

What about the celestial bodies like the Sun and Moon that shine much brighter than a first-magnitude star? This is where negative magnitude numbers come into use. The Sun and full Moon have magnitudes of −26.7 and −12.5, respectively. Venus, brightest of the planets, often shines as bright as magnitude −4.5. Jupiter is typically a magnitude −2.5 to −2.9 object. Sirius, the brightest of the nighttime stars, has a measured magnitude of −1.4.

....

JUPITER: LOOKING CLOSER

During the centuries that elapsed between the time of the ancient Greeks and the early years of the seventeenth century, almost nothing new was learned about Jupiter. That situation changed dramatically in 1610, when the pioneer astronomer Galileo Galilei trained his telescope on Jupiter. Galileo observed with his own eyes that Jupiter presented a distinct sphere, quite different from the stars that remained as sparkling points of light. Galileo also discovered four bright "stars" that orbit Jupiter the way the Moon orbits Earth. The discovery of moons circling another planet marked the beginning of the end for the longstanding Earth-centered model of the universe.

As the size and optical quality of the telescope improved, so did our knowledge of Jupiter. The banded nature of its atmosphere was discovered in the late 1600s, as was a huge oval-shaped disturbance called the Great Red Spot. The Great Red Spot served as a marker that allowed astronomers to determine Jupiter's rate of rotation. A value of slightly less than 10 hours explained the slightly flattened appearance of Jupiter's disk. The big planet was spinning so rapidly that it had developed an equatorial bulge.

....

Determining a Jovian Planet's Rotation

For astronomers in the seventeenth and eighteenth centuries, determining the rotation period of the terrestrial planet Mars was a relatively easy task. Pick a surface landmark. Time consecutive passages of the landmark across the center of the Martian disk and, *violà*, you have the planet's period of rotation.

Figuring out the rotation periods of the Jovian planets wasn't so clear-cut. Because Jupiter and Saturn have no solid surface, astronomers were forced to rely on readily recognizable disturbances in their gaseous atmospheres—Jupiter's Great Red Spot being a good example. However, atmospheric disturbances do not make reliable markers. Interactions with adjacent atmospheric features cause them to speed up or slow down. They become "moving targets."

Worse than that, the Jovian planets don't rotate as solid bodies the way the terrestrial planets do. Their spin rates vary with latitude. Some of the early figures for the rotation periods of Jupiter and Saturn were based on averages for different latitudes.

Uranus and Neptune presented a special problem for astronomers. They are so distant that finding observable atmospheric disturbances was all but impossible.

Our present-day knowledge about the rotation periods of the outer planets is based on the rotations of their cores. Since a planet's magnetic field will spin in unison with its core, astronomers can determine core rotation by analyzing the movement of the magnetic field, using interplanetary probes like *Voyager 2*. The rate of spin of the cores isn't that dramatically different from the spin of their atmospheres, and it's a more reliable figure.

....

How big is Jupiter? Astronomers know that the angular size of its disk is typically 45 **arcseconds** around the time of opposition. An arcsecond is an angular measure equal to 1/3,600 of a degree. On average, the full Moon has an angular diameter of 0.5 degree (1,800 arcseconds). Jupiter's disk appears to be about 40 times smaller than the Moon's, and roughly equal to the size a golf ball would appear if placed 0.2 kilometers (one-eighth of a mile) away. We know that Jupiter is bigger than either the Moon or a golf ball. If astronomers could determine Jupiter's distance, they could translate its apparent diameter in arcseconds into an actual diameter in kilometers or miles.

When the distance scale of the solar system was figured out and the gap that separates Jupiter from Earth calculated, astronomers at last calculated Jupiter's true diameter. What a huge planet it proved to be. Over 40 of our Moons laid end-to-end would barely cross its disk—and about 3.5 *trillion* golf balls!

Jupiter proved to be a massive world as well. Its gravitational pull on the Galilean moons provided the clues astronomers needed to determine

Jupiter's mass, using Newton's laws of gravity. In simple terms, Jupiter's mass could be calculated by knowing the distance of its moons from the planet's center and how fast they moved in their orbits. The greater the mass, the greater Jupiter's gravity pull, and the faster the moons would have to revolve to overcome its pull. Jupiter was found to have a mass equal to more than three hundred Earths. Jupiter's gravity, calculated at the height of its cloud tops, was found to be greater than 2.5 times Earth's. The only unimpressive statistic about Jupiter was its density, determined mathematically by dividing its mass by its volume. The big planet was only one-fourth as dense as Earth.

In the waning decades of the nineteenth century, astronomers began to turn their attention to the turbulent cloud bands encircling Jupiter and running parallel to its equator. Based on Jupiter's low density, they correctly surmised that Jupiter must be comprised of light gases. Early attempts at spectroscopic analysis proved inconclusive, leaving astronomers to guess at Jupiter's composition.

In 1892, nearly three centuries after Galileo had discovered Jupiter's four large moons, the American astronomer E. E. Barnard uncovered a fifth. The tiny moon, which appeared as a faint flicker of light in the eyepiece of the giant Lick Telescope in California, would be the last to be discovered visually. The human eye was about to be replaced by something far more sensitive.

JUPITER: A MODERN VIEW

The photographic plate, developed as an astronomical tool in the late 1800s, allowed astronomers to detect detail too faint to be seen by visually peering into the eyepiece. Between 1904 and 1951, seven new moons of Jupiter were discovered photographically.

In the 1930s, astronomers again used the spectroscope to analyze Jupiter's atmosphere, this time finding traces of methane and ammonia. In 1960, hydrogen was added to that list. Astronomers using the radio telescope in 1955 detected huge quantities of radio noise from Jupiter, some of it generated by Jupiter's magnetic field. The more astronomers looked at Jupiter, the more aware they became of its identity as a gas giant—a planet with no solid surface.

A veritable torrent of knowledge about Jupiter occurred with the arrival of the Space Age and the first flybys by interplanetary space probes. In 1973 and 1974, *Pioneer 10* and *11* flew past Jupiter, taking the first close-up images of the planet and its turbulent atmosphere. *Voyagers 1* and *2* followed in 1979, discovering Jupiter's rings, and providing dramatic close-up views of the Galilean moons. A highlight of the *Voyager* missions was the discovery of active volcanoes on Jupiter's moon Io, forever erasing the conception of planetary moons being dead worlds like ours. The *Galileo* mission to Jupiter was the first orbital probe. During an eight-year odyssey beginning with *Galileo*'s arrival at Jupiter in 1995, the *Galileo* craft imaged Jupiter cloud patterns, analyzed its magnetosphere, and mapped its moons. A probe was dropped into

Jupiter's atmosphere to sample the composition of the gases in its upper levels.

Today, astronomers continue to study Jupiter from Earth, now using telescopes and equipment much larger and more sensitive than the ones used in previous decades. The orbiting Hubble Space Telescope will continue to augment ground-based observations.

Future exploration of Jupiter will be undertaken by larger and more technologically advanced ground-based telescopes, some perhaps even stationed on the Moon. The Hubble Space Telescope will be joined by the James Webb Telescope, scheduled for launch in 2013. Future robotic missions to Jupiter will include an orbiter (the Juno Mission, scheduled for a 2011 launch), followed by soft landings on the moons and, ultimately, a manned mission.

KING OF THE PLANETS

Jupiter is big—period. Statistics on this giant and its equally impressive moons read like a commentary from Ripley's "Believe It or Not":

- With an equatorial diameter of 142,984 kilometers (88,848 miles), Jupiter is as wide as 11 Earths placed side by side.
- Were it hollowed out, Jupiter could accommodate over 1,300 Earth-sized planets.
- If we could place Jupiter on one side of a giant scale, we would need more than three hundred Earths on the other side to balance it out.
- Jupiter contains over twice the mass of the other planets in the solar system combined.
- The radiation belt surrounding Jupiter is so lethal that it would kill an astronaut almost instantly.
- Jupiter's magnetic field is nearly 20,000 times as strong as Earth's.
- The magnetosphere created by Jupiter's magnetic field is the largest entity in the solar system. If it gave off visible light, it would appear twice as large as the full Moon in our night sky.
- The Great Red Spot, a huge storm in Jupiter's atmosphere, is two times as large as the Earth.
- Jupiter rotates once in just 9 hours, 56 minutes. At its equator, this produces a rotational speed of 35,000 kilometers (21,000 miles) per hour.
- Jupiter is a miniature solar system, attended by 63 moons.
- Jupiter's largest moon, Ganymede, is bigger than the planet Mercury.

Yes, the late Mr. Ripley would have had a lot of fun describing this planetary colossus. In fact, Jupiter is about as big as a planet can get. Adding more material would increase Jupiter's gravity, causing it to compress and maintain its current diameter. This leads to an interesting question. How close did Jupiter come to being a star?

A FAILED STAR?

In the simplest of definitions, a star is a huge, gaseous body that radiates into space vast quantities of energy—in particular, heat and light. A planet, on the other hand, is smaller and releases primarily the energy it acquires from its parent star. This is not the case with Jupiter, which gives off about twice the energy it receives from the Sun. Considering its immense size, gaseous composition, and ability to generate its own internal heat, is it realistic to call Jupiter a "failed star"? Not really!

The excess energy comes not from the fusion of hydrogen into helium, as is the case with a star like the Sun, but rather from heat generated by Jupiter's gradual gravitational collapse and the "raining out" of helium deep in Jupiter's atmosphere. Jupiter's core is undeniably hot—perhaps two to four times the temperature of the Sun's surface. But it's nowhere near equal to the nuclear furnace that exists at the Sun's core, where a temperature in excess of 20 million degrees Celsius and pressure millions of times that exerted by our atmosphere generate nuclear fusion.

If we were to add mass to Jupiter, the increased gravity would compress the core, causing its temperature and pressure to increase. But to reach the conditions that would spark nuclear fusion and transform this huge planet into a small star would require an additional 75 to 80 Jupiter masses. To become a "**brown dwarf**," an object intermediate between planet and star, would entail the addition of at least a dozen Jupiter masses. The King of the Planets is hardly stellar!

ORBIT AND ROTATION

Fifth planet outward from the Sun, Jupiter orbits the Sun at an average distance of 778.6 million kilometers (483.8 million miles). Astronomers sometimes express distances in the solar system using a yardstick called the **astronomical unit** (AU). Because one astronomical unit equals the average distance of the Earth from the Sun (150 million kilometers, or 93 million miles), it's a convenient way to compare a planet's distance from the Sun with Earth's. Since Jupiter is 5.2 times farther from the Sun than Earth, we say it is 5.2 astronomical units from the Sun.

Jupiter's average orbital speed is 13 kilometers (8 miles) per second. At that pace, Jupiter could travel cross-country from New York to Los Angeles in a little more than five minutes. Even so, Jupiter takes nearly 12 Earth years to complete a full orbit of the Sun. An average human life span would equal about six Jupiter years.

Oddly enough, our solar system's largest planet also spins the fastest, completing a rotation in a mere 9 hours, 56 minutes—2.5 Jupiter days for every one of ours. This rapid rotation creates a noticeable bulge at Jupiter's equator, where the planet is racing around at a 35,000 kilometer (21,000 mile) per hour clip. Because Jupiter is an **oblate spheroid**, the distance between its poles is over 9,200 kilometers (5,700 miles), less than its equatorial diameter.

Voyager 2 Gives Jupiter a Shove

Voyager 2 was able to complete the Grand Tour of all four outer planets within 12 years of its launch because of the gravitational boost it got as it passed each one. You may be surprised to learn that the car-sized craft gave each of these giants a push in return.

Whenever you jump, the Earth's gravity pulls you back to the ground. Not only does the Earth exert its gravitational pull on you, but also you exert a tiny gravitational pull on the Earth. As you fall downward, the Earth actually falls upward to meet you. The amount the Earth moves is infinitesimally small, however.

As *Voyager 2* plummeted towards Jupiter, the huge planet was moving an almost immeasurably small distance toward the speeding craft. *Voyager*'s trajectory was designed to miss Jupiter and zip onward at an increased speed toward Saturn. Upon leaving Jupiter, *Voyager 2* had increased its speed by a whopping 57,600 kilometers (35,700 miles) per hour. In the process, *Voyager 2* slowed Jupiter, whose mass equals 318 Earths. As a result of its encounter with *Voyager 2*, Jupiter will travel about 30 centimeters (one foot) slower in its orbit—over the next 1 trillion years.

AN ATMOSPHERIC MAELSTROM

Jupiter looks pretty peaceful when viewed through the eyepiece of a telescope perched hundreds of millions of kilometers away. But looks can be deceiving. Its atmosphere, driven by internal heat and separated into alternating light and dark bands by its rapid rotation, is a veritable maelstrom. Winds exceeding 640 kilometers (400 miles) per hour buffet the planet's upper atmosphere, while friction between adjacent belts generates Earth-sized cyclonic storms.

Long before the two *Voyagers* (and *Pioneer 10* and *11* before that) sent back photos and data about Jupiter's atmosphere, astronomers were aware of its turbulent nature. Mars displayed distinct surface markings that were occasionally obscured by enormous dust storms. Jupiter, on the other hand, appeared perpetually overlain with a dense, banded atmosphere interspersed with temporary light-hued ovals and punctuated by the larger and more permanent Great Red Spot.

The bands, readily visible in telescopes, appear as alternate light and dark stripes running east to west across Jupiter's disk. While our weather is powered by the Sun, which creates areas of differing temperatures, Jupiter's

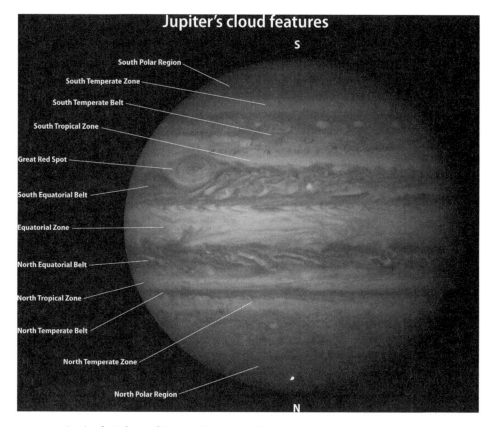

Figure 4.2 Jupiter's Belts and Zones. Courtesy of *Astronomy* magazine.

weather is created by upwellings of heated gases from deep within the planet. The lighter bands, called **zones**, are created by vertically rising gases from deep within Jupiter. The updrafts result in high-altitude clouds of frozen ammonia. Winds within these high-pressure zones travel in an easterly direction. The reddish-brown bands, called **belts**, are much lower in elevation and pressure, move in a westerly direction, and are colored by clouds of ammonium hydrosulfide. Two belts, the North and South Equatorial Belts, are readily observable in small backyard telescopes. Larger telescopes reveal several more belts. Unlike atmospheric patterns on Earth, which can break up as they pass over mountainous terrain, Jupiter's belts and zones appear to be rather constant.

A STORM BIGGER THAN EARTH

A maelstrom is defined as a violent or turbulent situation. Few words more aptly describe the conditions that prevail within Jupiter's atmosphere, particularly Jupiter's most celebrated feature, the Great Red Spot. The earliest reliable observation of this huge oval patch is credited to Giovanni Cassini of Italy, who reported its existence in 1664. The mysterious object was observed for the next half century. After a 118-year gap, it was again observed in 1830. The Great Red Spot has been visible ever since, although its appearance varies. At times, it appears brick red and is readily seen in small backyard telescopes. At other times, it fades to light pink and is barely visible as a dent in Jupiter's South Equatorial Belt.

The nature of the Great Red Spot was the subject of much debate, especially when astronomers realized that it drifts back and forth in longitude. Early ideas about its nature centered on perceptions of its being a floating mass of some lightweight solid substance or perhaps material spewed by an unseen volcano.

Close-up photos provided by the *Voyagers* showed that the Great Red Spot is a vast storm—an anticyclone (high-pressure system) rotating counterclockwise in Jupiter's southern hemisphere. And what a storm it is! An oval 25,000 kilometers (15,500 miles) across and half as wide, it could hold two Earths side by side. Winds whip around the edge of the Great Red Spot at speeds approaching 360 kilometers (225 miles) per hour—twice as fast as a typical terrestrial hurricane. Clouds at the top of the Great Red Spot tower above the cloud tops in adjacent parts of the atmosphere.

How long will the Great Red Spot last? Hurricanes on Earth rarely survive more than a few weeks, due to friction with the Earth's hilly surface. Since Jupiter lacks such a surface, the Great Red Spot could rage unabated for centuries. The Red Spot has shrunk somewhat since the time of its discovery, but astronomers, professional and amateur, can expect to enjoy this spectacle for many decades to come.

RED SPOT JR.

The Great Red Spot isn't the only storm in Jupiter's atmosphere. It's just the largest and longest-lasting. Over the years, other Jovian storms, all basically oval in shape, have been studied by professional astronomers at major observatories. A few have been known to generate powerful lightning bolts, some with hundreds of times the jolt of an earthly lightning bolt. Some of Jupiter's storms are so large that amateur astronomers using moderate-sized backyard telescopes can observe them. Such is the case with a recent storm nicknamed "Red Spot Jr."

In 2000, three small white spots in Jupiter's atmosphere collided and merged into one large white oval about the size of the Earth. Astronomers assigned it the rather mundane name "Oval BA." The storm remained white for several years, but near the end of 2005, as Jupiter approached conjunction with the Sun, began to take on a brownish hue. When Jupiter emerged from the Sun's glare in February 2006, amateur astronomer Christopher Go, of the Philippines, photographed it with an 11-inch telescope and CCD camera—equipment as sophisticated as that used by professional astronomers a few decades earlier.

In Go's photograph, Oval BA was red! Astronomers around the world have been studying Red Jr. since. How long will this storm last? Will it collide and merge with the Great Red Spot? With no clear idea, we can only wait and watch.

JUPITER TAKES A HIT!

Although much of the planet-building in our solar system occurred in its first few hundred million years, the planets continue to grow today. In a single day, the Earth accumulates tens of thousands of tons of material from micrometeorites that enter our atmosphere and drift lazily to the ground. On rare occasions, a large meteoroid or comet will impact the Earth.

Jupiter, with its far-reaching gravity field, is especially adept at gathering bodies that stray too close. In July 1994, a historically unparalleled event occurred with Jupiter at center stage. Astronomers around the world watched with baited breath as the remnants of a comet named Shoemaker-Levy 9 crashed into Jupiter.

Sixteen months earlier, astronomers Eugene and Carolyn Shoemaker and David H. Levy discovered the comet on a photographic plate made with a 0.4-meter (18-inch) telescope located at the Mt. Palomar Observatory. Orbital calculations showed that the comet had recently passed near Jupiter and was pulled into a fatal path that would bring it crashing into the planet. Close-up photos of Shoemaker-Levy 9 showed that it was not one, but possibly 21 distinct bodies strung out like cosmic pearls. Apparently, the comet had broken apart during its earlier Jupiter encounter.

As the moment of first impact approached, telescopes around the world were pointed toward Jupiter. Above the Earth's atmosphere, the Hubble Space Telescope was at the ready, as was the Jupiter-bound *Galileo* probe. Although the comet fragments landed on Jupiter's far side, the planet's rapid rotation quickly brought the impact sites into view.

One dramatic series of photos taken by the Keck Telescope in Hawaii showed a brilliant fireball thrusting upward from Jupiter's limb, then just as suddenly dropping back out of sight. Moments later, a blackened scar rotated into view. For several days, astronomers worldwide marveled at the sight of a series of dark, sooty-looking blotches stretched out across Jupiter's disk. These Earth-sized scars were readily visible for weeks afterward, even with small backyard telescopes.

Astronomers learned much from the Shoemaker-Levy 9/Jupiter encounter. As each cometary fragment plunged into Jupiter's atmosphere and exploded, it dredged up material from lower levels of the atmosphere, which astronomers could analyze with the spectroscope. The Shoemaker-Levy 9 impacts also gave astronomers dramatic insight into how Jupiter has accrued matter from the solar system. The events of July 1994 were merely the latest in what must have been countless cometary impacts in Jupiter's long history—each adding material to the big planet.

The Shoemaker-Levy 9/Jupiter encounter was a sobering reminder that the planets are vulnerable to a comet impact. The chunks of Shoemaker-Levy 9 that plowed into Jupiter left blackened Earth-sized smudges in Jupiter's atmosphere that dissipated within weeks. Within a year, there was no evidence of the event that had captured the attention of astronomers around the world. And yet, had just one of those cometary chunks hit Earth, the result might have been a global disaster.

JUPITER'S INTERNAL STRUCTURE

An analysis of Jupiter's atmosphere with the spectroscope reveals that it is comprised primarily of molecular hydrogen (H_2, 89.8 percent) and helium (He, 10.2 percent) by volume—a composition similar to the Sun and, quite likely, the nebula that gave birth to the solar system. Such gases are color-less, so the yellows, blues, reds, and browns seen in Jupiter's atmosphere must arise from the trace amounts of methane (CH_4), ammonia (NH_3), and water (H_2O) that have also been detected.

What we see when we observe Jupiter through the telescope is the top of its atmosphere. The highest part is found in the Great Red Spot. Tempera-tures here are about $-145°C$ ($-230°F$). Jupiter's pale-yellow zones exist at slightly lower altitudes, and are made up of ammonia ice crystals. Further down are the reddish-brown ammonium hydrosulfide [$(NH_4)SH$]crystals found in Jupiter's belts. As we move deeper into Jupiter's atmosphere, tem-peratures and pressure increase. Below the ammonium hydrosulfide layer, we find water ice, then a water/ammonia fog.

For all its superlatives, Jupiter falls behind our Earth in one key trait, den-sity. That's because our Earth is comprised of heavier materials—rocky sili-cates and metals like iron and nickel. Jupiter, meanwhile, is made up of much lighter substances, like hydrogen and helium. Jupiter's overall density is just 1.3 times greater than water and less than one-fourth that of our Earth.

As noted in Chapter 1, the term *gas giant* to describe Jupiter and the remaining outer planets is misleading. There is a depth in Jupiter's atmos-phere where the pressure compresses the hydrogen in Jupiter's atmosphere into a liquid state. This occurs approximately 4,000 kilometers (2,500 miles) deep. While Earth's atmosphere extends upward a few dozen kilometers at best, Jupiter's atmosphere is a full Moon diameter in depth.

The liquid molecular hydrogen zone continues downward for about 7,000 kilometers—about one-third of the way from the cloud tops to the planet's center. At pressures *4 million* times greater than our atmospheric pressure at sea level and a temperature comparable to that which exists at the Sun's surface, hydrogen now acts like a molten metal, able to conduct electricity. It is here that Jupiter's magnetic field is produced.

This ocean of liquid metallic hydrogen continues all the way to Jupiter's core—a superhot, slushy mixture of ices, silicate rock, and metals estimated at anywhere from 3 to 20 times as massive as the Earth. Such a wide range of masses demonstrates how little astronomers truly know about what lies deep beneath Jupiter's turbulent atmosphere.

How Do We Know What's Inside a Planet?

How can astronomers say that Jupiter has a liquid metallic layer buried a certain number of kilometers beneath its surface? They can't! Our models of the internal structure and composition of the outer planets Jupiter, Saturn, Uranus, and Neptune are a mixture of painstaking data collection and analysis, computer modeling, and educated guesswork.

The process starts with the determination of the planet's overall density. First, astronomers calculate its volume—a simple mathematical process once you know its diameter. Next, they establish the planet's mass. Not as straightforward as volume, a planet's mass is determined by studying its gravitational effect on orbiting moons and applying Newton's laws of gravity. Once a planet's volume and mass are known, astronomers compute its average density by dividing its mass by its volume. The final answer is expressed in grams per cubic centimeter (g/cm^3).

Jupiter's average density has been figured at 1.33 g/cm^3. This is only slightly greater than water and one-fourth Earth's. Remember, it's an *average* figure. The density in Jupiter's upper layers will be much less than near the highly compressed core. At least we can rule out the idea that Jupiter is made up entirely of gases (densities far less than 1.0 g/cm^3) or metal (densities of 7 g/cm^3 and up).

What elements might be present in Jupiter? Here's where the spectroscope takes over. Spectroscopes attached to earthbound telescopes and those mounted on interplanetary probes analyze the spectrum of Jupiter's atmosphere, looking for the spectral fingerprints of elements and compounds. Since these are the materials found at or near Jupiter's cloud tops, we make an educated guess as to what materials lie closer to the core. The metals and silicates associated with the meteoric and cometary debris that rains down on Jupiter probably sink towards the core.

Like throwing ingredients in a blender, astronomers program the necessary information in a computer. With luck, they'll wind up with a logical model of composition, temperature, and pressure from cloud top to core. Whatever the model, it will last until new data from ground-based instruments or an orbiting space probe makes it obsolete.

Where Is Jupiter's "Surface?"

The diameter of a terrestrial planet like Earth is basically defined as the distance from one side of its rocky surface at the equator, through the center of the planet, to the other side. What happens when we try to determine the diameter of one of the Jovian planets, where no solid surface is available? In the case of Jupiter and the other outer planets, the traditional starting point has been the top of the visible cloud layers.

JUPITER'S MAGNETOSPHERE

If you've ever sprinkled iron filings on a piece of paper that was covering a bar magnet, you saw a pretty graphic display of a magnetic field. It's created by the atomic structure of the metals in the magnet. The filings line up along the magnetic field lines from one pole of the magnet to the other.

A magnetic field can also be generated by the movement of electrons in an electric current. This is the principle by which electromagnets work, and it's the way the planets generate magnetic fields. We know that the Earth has a magnetic field, by its effect on compasses. Earth's magnetic field is generated by the movement of electrons in its molten nickel-iron outer core.

Jupiter possesses the strongest magnetic field of all the planets, measured to be 20,000 times stronger than Earth's. Astronomers aren't quite sure of the source, but it's believed to originate in the liquid metallic hydrogen zone one-third of the way toward the core. As the core rotates, it sets up a current of electrons that generates a powerful magnetic field. Jupiter's magnetic field, like ours, is tipped relative to its axis of rotation. In Jupiter's case, the tilt amounts to 10 degrees.

As is the case with Earth and the other planets that have magnetic fields, this one captures particles from space, especially the charged particles like electrons and protons streaming out from the Sun (the **solar wind**). These particles, thus trapped in the magnetic field, form a planet's **magneto-sphere**. The solar wind actually compresses Jupiter's magnetosphere on the side facing the Sun and stretches it outward on the opposite side, creating a teardrop shape. The side of Jupiter's magnetosphere facing sunward extends 3 to 7 million kilometers (1.9 to 4.3 million miles), depending on the strength of the solar wind. In the other direction, it stretches outward like a huge wind sock to the orbit of Saturn 750 kilometers (466 million miles) away. In fact, Jupiter's magnetosphere is considered the largest structure in the solar system. If it glowed in visible wavelengths, it would appear twice as large as the full Moon in our night sky.

As Jupiter rotates, its magnetic field rotates with it, sweeping across the three inner Galilean moons. Io is especially affected. As Jupiter's magnetic field lines sweep across Io, they remove about 1,000 kilograms (over one ton) of material every second. This material, comprised mainly of sulfur and oxygen **ions**, forms a doughnut-shaped **plasma torus** that surrounds Jupiter along Io's orbit. It's a zone of intense radiation.

Some ionized particles from Io follow Jupiter's magnetic field lines directly toward the planet's magnetic poles, where they produce intense **auroral** displays. The electrical current associated with this stream of ions generates powerful lightning bolts.

BAD NEWS FOR FUTURE JUPITER EXPLORERS

Imagine that you've been asked to be part of a team planning for the first human expedition to Jupiter. Your assignment is to select an appropriate place to land and set up a base of operations. You already know that a landing on Jupiter is a virtual impossibility because Jupiter lacks a solid surface to park a spaceship.

How about the Galilean moons? The inner three are immersed in the deadly radiation belt associated with Jupiter's magnetosphere. That leaves the outermost moon, Callisto, as the only safe landing spot for a Jovian expedition.

JUPITER, THE RINGED PLANET

The discovery by the *Voyagers* of a faint ring around Jupiter didn't come as a complete surprise. Saturn's rings had been known for centuries, and a ring around Uranus was discovered in 1977, the year the *Voyagers* were launched. In fact, both *Voyagers* were specifically programmed to search for a Jovian ring system. To that end, they were successful.

Jupiter's ring system is extremely faint, visible only by backlit sunlight. As *Voyager 1* and *2* passed Jupiter, they turned sunward for a look. Sunlight scattered as it passed through the rings betrayed their presence as a bright streak against the black background of space. The effect was similar to what you see when sunlight passes through a car windshield and reveals streaks in the glass.

Much dimmer and less extensive than Saturn's ring system, Jupiter's rings are comprised of three main parts. A main ring about 7,000 kilometers (4,300 miles) wide and less than 20 kilometers (12.5 miles) thick is situated between a halo-shaped ring that extends all the way to Jupiter's cloud tops and an outer gossamer ring that reaches outward to the orbits of Jupiter's moons Amalthea and Themis. The rings are likely comprised of extremely fine dust particles, possibly originating from material cast off by these two moons and two others that lie closer to Jupiter.

LIFE ON JUPITER?

In 1870, the British astronomer Richard A. Proctor wrote *Other Worlds than Ours*. In it, he discussed the nature of any life that might exist on Jupiter. Like many of the astronomers of his day, Proctor believed that Jupiter had a solid, inhabitable surface beneath its dense banded atmosphere. He wrote, "The grandeur of his (Jupiter's) orb naturally suggests, at first sight, the idea of beings far exceeding, both in might and bulk, those which live upon the earth." Proctor then related the suggestion of another astronomer who asserted that the Jovians must be "fourteen feet high by eye-measurement." This calculation was based on the assumption that inhabitants of Jupiter might have a body structure similar and proportionate to ours. Since Jupiter receives much less light from the Sun than we do on Earth, larger eyes (and therefore a proportionately larger body) would be a requisite for survival. Proctor then suggested an alternate Jovian being, small in height and weight, to compensate for Jupiter's strong gravity.

We might find such descriptions of life on Jupiter as laughable and naive, but they were based on the scant knowledge of the planet that astronomers had to work with over a century ago. What would a hypothetical Jovian creature look like from the viewpoint of a modern-day astronomer?

To begin with, *inhabitus Joviensis* wouldn't have arms and legs. Such appendages would have little use on a planet with no solid surface. Since conditions in the cold, turbulent upper atmosphere of Jupiter might be too harsh for life, we might have to look deeper, where a calmer, warmer environment exists. Pressures here would be far greater than what we experience on the Earth's surface, compressing the gases into a slushy state. The small quantity of sunlight that reaches Jupiter's cloud tops many kilometers above would be all but absorbed at this depth. Jovian life forms might take on the appearance of the bizarre creatures that exist in the dark, highly pressured depths of our oceans.

One drawback to the existence of Jovian life is the presence of powerful upwellings of material rising from Jupiter's interior. Such upwellings might rapidly force *inhabitus Joviensis* from its comfort zone to high altitudes. It might not survive when suddenly subjected to the colder temperatures and sudden drop in pressure. Only a future expedition to Jupiter and a probe sent to the depths of the planet's atmosphere will answer the question, "Is Jupiter an abode for extraterrestrial life?"

SEE FOR YOURSELF

You can always get a good idea of Jupiter's appearance simply by looking at the photos in this book or by perusing the pages of a monthly magazine like *Astronomy* or *Sky and Telescope*. The Internet is crammed with sites devoted to the big planet. But if you want a first-hand look at Jupiter, why not go outside and capture a real, honest-to-goodness view?

It helps to know where Jupiter is in the night sky at the time you want to view it. You'll find this information in a monthly astronomy magazine or on an Internet site (you can refer to Web Sites at the end of this chapter or Appendix E for examples). Once you know the general area of the sky where to look, you'll have no trouble locating Jupiter. At an average visual magnitude of −2.5, Jupiter easily outshines all of the nighttime stars. The only starlike object that's brighter is the planet Venus. When Jupiter is near opposition, it's about 630,000,000 kilometers (391,500,000 miles) away. At that distance, light from the planet takes about 35 minutes to reach your eyes.

If Jupiter is currently in the morning sky or, worse yet, at conjunction hiding behind the Sun, be patient. In a matter of months, it will again grace the evening sky, commanding the attention of backyard sky gazers.

Do you prefer an up-close-and-personal view? Don't worry. You don't have to be a professional astronomer with access to space probes like *Voyager*, huge ground telescopes like the Keck in Hawaii, or the Hubble Space Telescope. You can "Jupiter Watch" from the comfort of your backyard with nothing fancier than a good pair of binoculars or a small telescope.

Jupiter is so big that, even though it's more than two thousand times farther away from us than the Moon, we can still observe it with binoculars. A typical pair of binoculars will enlarge Jupiter 6–10 times. This isn't enough magnification to reveal intricate detail, but it's adequate to make out the planet's disk and capture a few, if not all, of the four brightest moons. Because their orbits run edge-on to our line of sight, the Galilean moons appear to be strung out to Jupiter's sides.

You'll have no trouble spotting the Galilean moons with a small telescope and a magnification of just 30X. What you see the first time you train a telescope on Jupiter will depend on the positions of the moons at that moment. All four may be arranged to one side of Jupiter. You may see a combination of three on a side, one on the other, or two on each side. Not all of the moons may be visible. Now and then, one or two moons may be hiding behind Jupiter. What you see that first night will be different the next. The Galilean moons move rapidly, with periods ranging from just under two days for Io, nearest of the four moons to Jupiter, to a little over two weeks for distant Callisto.

The major astronomy magazines publish tables showing the positions of Jupiter's moons for any date and time of the month. Especially interesting are eclipses of Jupiter's moons as they disappear into the big planet's shadow. Equally fascinating are transits of the moons in front of Jupiter. A transiting moon may be invisible, camouflaged against Jupiter, but its shadow on Jupiter will appear as a slowly moving black dot.

Besides the four Galilean moons, a small backyard telescope will also reveal Jupiter's slightly flattened, banded disk. Overall, Jupiter's disk is made up of alternate pale-yellow zones and grayish to reddish-brown belts that run parallel to the equator. The two most prominent belts, the North and South Equatorial Belts, are easily seen in small telescopes. With a medium-sized telescope—one whose objective lens or mirror diameter is 10 to 15 centimeters (4 to 6 inches)—and a magnification of 100X or more, you might be able to pick out more bands, as well as the Great Red Spot if it's currently on the side of Jupiter facing Earth. If not, wait a few hours. Jupiter's rapid rotation will bring the Great Red Spot into view in a matter of hours.

Experienced amateur astronomers, using large telescopes that magnify up to 300X, can see intricate detail in the cloud bands and follow the development of atmospheric disturbances that are smaller than the Red Spot. Those who attach CCD cameras to their telescope are able to take images of Jupiter that rival those taken by the world's major observatories just a few decades ago. The same technology that helps professional astronomers is available in scaled-down versions to amateur astronomers.

WEB SITES

http://www.nineplanets.org/jupiter.html.

    Bill Arnett's popular Nine (8) Planets Web site. Information-filled look at Jupiter.

http://sse.jpl.nasa.gov/planets/profile.cfm?Object=Jupiter.

    Facts about Jupiter, gleaned from various NASA missions.

http://www.astronomycast.com/astronomy/episode-56-jupiter.

    The written script of a podcast on Jupiter done by Fraser Cain and Pamela Gay. Contains links to the actual podcast and other Web sites that provide information about Jupiter.

5

Jupiter's Moons: A Solar System in Miniature

FOUR STARS FOR A PRINCE

On the evening of January 7, 1610, the Italian astronomer Galileo Galilei trained a crudely crafted refracting telescope on Jupiter. To his surprise, he saw that the planet was closely attended by three bright stars. Their straight-line arrangement aroused his curiosity, and he began observing Jupiter each clear night. Galileo soon realized that this stellar trio, which was later joined by a fourth discovered on January 13, was in orbit around Jupiter. He named these wanderers the Medician stars, in honor of his patron the Grand Duke Cosimo de' Medici, a prominent member of the Medici family of Florence.

Apparently, Galileo wasn't the only one peering through a telescope at Jupiter. In 1614, the German astronomer Simon Marius published the book *Mundus Iovialus* [The Jovian World], in which he asserted that he had discovered Jupiter's moons weeks before Galileo. Because *Mundus Iovialus* lacked observational sketches to prove Marius's claim (and because Marius's reputation had been stained a decade earlier when a student he had mentored plagiarized one of Galileo's works), he wasn't taken seriously. However, his suggestion that the Jovian satellites be named Io, Europa, Ganymede, and Callisto was ultimately accepted. Marius may not have stolen Jupiter's moons from Galileo, but he stole their names from the Prince of Florence.

Astronomers today refer to the quartet as the Galilean satellites, a name suggested by Galileo's contemporary, the German astronomer/mathematician

Galileo Galilei (1564–1642)

The name Galileo is synonymous with astronomy, but he also made his mark in physics and mathematics. Born in Pisa, the young Galileo aspired to the priesthood, but later attended school to study medicine. His interests turned to science and mathematics and, in 1589, Galileo was appointed to the Chair of Mathematics at the University of Pisa.

Figure 5.1 Frontispiece, Portrait of Galileo, The Assayer, 1623. Images copyright History of Science Collections, University of Oklahoma Libraries.

What earned Galileo a place in history were his exploits with the telescope. Once he learned of this remarkable new invention that could almost magically make distant objects appear closer, Galileo set about to construct one of his own. Between 1609 and 1610, Galileo turned his telescope heavenward and made a series of astronomical discoveries that were to revolutionize the way humans regarded the cosmos. His reports of craters and mountains on the Moon and dark blotches (sunspots) on the Sun countered the popular belief that both were perfect heavenly spheres. Observations of Venus showed that the planet undergoes a series of Moon-like phases while dramatically changing its apparent size. This was possible only if Venus circled the Sun and not the Earth as was commonly believed. The four moons observed by Galileo to be orbiting Jupiter also helped to erode the idea that all things heavenly circled the Earth.

Galileo's published reports of his discoveries and his staunch support of a Sun-centered universe placed him in direct conflict with the Roman Catholic Church, which traditionally promoted an Earth-centered universe. He was ultimately brought before the Inquisition, where he was ordered to recant his beliefs and was barred from ever again teaching them. The final years of his life were spent under house arrest.

••

Johannes Kepler. As was mentioned earlier in this book, the Galilean moons are of historical significance. Galileo's announcement of their existence shook the very foundation of human thought, for in those days it was commonly believed that the Earth stood alone and fixed at the center of the universe. To find celestial bodies orbiting something other than Earth was unthinkable. The rethinking process that would ultimately depose the Earth as the center of the universe and replace it with the Sun would take decades.

A SOLAR SYSTEM IN MINIATURE

The King of the Planets is attended by a most impressive court. As of 2008, no fewer than 63 moons are known to orbit Jupiter. When we look at this huge planet and its system of satellites, it's as though we are regarding a miniature version of the solar system.

For nearly three centuries, the Galilean satellites were the only known bodies orbiting Jupiter. The first indication that Jupiter might have other satellites came in 1892, when the American astronomer Edward Emerson Barnard (1857–1923) discovered Jupiter's fifth moon, Amalthea. Amalthea would be the last planetary moon to be detected by direct visual observation. Soon afterward, the photographic plate with its greater sensitivity to faint light would replace the human eye as the primary tool in the search for planetary satellites.

During the first half of the twentieth century, seven more Jovian moons were discovered: Himalia (1904), Elara (1905), Pasiphae (1908), Sinope (1914), Lysithea (1938), Carme (1938), and Ananke (1951). In 1974, Charles Kowal, working at the Mt. Palomar Observatory, captured images of a thirteenth Jovian moon, later called Leda. The next year, Kowal, assisted by Elizabeth Roemer, discovered a fourteenth moon. It was provisionally designated as S/1975 J 1. In discovery nomenclature, the "S" stands for satellite, "1975" the year of discovery, "J" for Jupiter, and the number "1" for the order of discovery for that year. Due to insufficient observations, the orbit of this moon could not be calculated and it was subsequently lost.

The arrival of the Space Age during the latter half of the twentieth century ushered in an era of robotic flybys of the planets. During its Jupiter encounter in 1979, *Voyager 1* imaged three new moons, Metis, Adrastea, and Thebe, all orbiting close to Jupiter.

The *Voyagers* did much more than just discover new Jovian moons. Prior to the *Voyager* flybys of Jupiter, moons were considered to be second-class citizens of our solar system—carbon copies of our own Moon. They were regarded as little more than overgrown, crater-pocked rocks orbiting the planets. To be sure, Io was noted for its distinct yellowish color, and Saturn's largest moon, Titan, was known to possess a substantial atmosphere. Other than those interesting tidbits, the planetary moons merited little more than a footnote in textbooks dealing with the solar system.

In 1979, Jupiter's moons leaped from back page footnotes to front page news headlines. Close-up images of the Galilean satellites relayed back to Earth by the *Voyager* spacecraft revealed them to be astoundingly unique, alien worlds. Io, in particular, was anything but a dead chunk of rock. Its surface was littered with active volcanoes. The remaining three were decidedly non-lunar in appearance. Jupiter wasn't surrounded by a bunch of lifeless rocks; it was the focus of four worlds as varied as any of the Sun's planets.

At the close of the twentieth century and during the early years of the twenty-first, several teams of astronomers using state-of-the-art CCD cameras mounted on some of the large telescopes atop Mauna Kea, Hawaii, added nearly four dozen moons to Jupiter's family, including Kowal and Roemer's "lost" S/1975 J 1, which was renamed Themisto.

THE NATURE OF JUPITER'S MOONS— AN ALL-OR-NOTHING SCENARIO

Where size is concerned, Jupiter's moons present us with an all-or-nothing scenario. The four moons discovered by Galileo in 1610—Io, Europa, Ganymede, and Callisto—are giants and would enjoy planetary status were they in individual orbits around the Sun. Of the remaining Jovian satellites, only 10 have diameters greater than 10 kilometers (6 miles). The largest of these, the football-shaped Amalthea (am-al-THEE-uh), spans just 250 kilometers (155 miles) along its greatest dimension. The majority of Jupiter's moons are mountain-sized chunks of rock and ice, orbiting far from the planet in highly inclined, retrograde (backward) orbits. They are most likely asteroids and comets that strayed too close to Jupiter and were captured by its gravity.

For the sake of simplicity, astronomers have grouped Jupiter's moons according to their orbital characteristics. From Jupiter and moving outward, they include the Amalthea Group, the Galilean Satellites, and the Himalia, Ananke, Carme, and Parsiphae Groups. Two moons, Themisto (located in a gap between the Galilean satellites and the Himalia Group) and Carpo (between the Himalia and Ananke Groups), have unique orbital characteristics and are not associated with any group.

THE AMALTHEA GROUP: JUPITER'S RING KEEPERS

Jupiter's moons are either gargantuan or lilliputian. Consider the football-shaped Amalthea, largest of Jupiter's non-Galilean satellites. You would need to line up 12 Amaltheas to cover a distance equal to the diameter of Europa, smallest of the Galilean moons. Amalthea is so small (approximately the size of the state of Vermont) that it wasn't discovered until nearly three centuries after the Galilean moons. It was observed in 1892 by the keen-eyed astronomer Edward Emerson Barnard, who spotted it with the 91-centimeter (36-inch) refracting telescope at the Lick Observatory in California.

One of Jupiter's closest moons, Amalthea zips around its home planet twice in one Earth day at an average distance (from Jupiter's center) of 181,300 kilometers (112,660 miles)—about half the Earth-Moon distance. If you were to hold a basketball 30 centimeters (one foot) in front of your face, you would have an idea how large Jupiter looms in the skies of Amalthea.

For the first 80 years following its discovery, little was known about Amalthea, except for its approximate size and orbit. That situation changed in 1979 with the arrival of the *Voyager* missions. The first discovery was that Amalthea is locked into a **synchronous orbit** with Jupiter, its long axis perpetually pointed toward the planet. Color photos of Amalthea showed that it's the reddest object in the solar system. The ruddy hues are likely a result of sulfurous material belched into space by Io's volcanoes and swept up by Amalthea. *Voyager* data also indicated that Amalthea releases more heat energy into space than it receives from the Sun. This is either due to the effects of orbiting within Jupiter's magnetosphere or tidal forces from the big planet.

During its 1979 flyby of Jupiter, *Voyager 1* imaged three new moons: Metis (MEE-tis), Adrastea (a-DRAS-tee-uh), and Thebe (THEE-bee)—in orbit near Amalthea. Like Amalthea, they have direct orbits, meaning they circle Jupiter in the same direction the big planet rotates. All are small and irregular in shape, with diameters between 20 and 100 kilometers (12 and 62 miles). The orbits of Metis and Adrastea lie within Jupiter's **Roche Limit**—a zone so close to a planet that the planet's gravity field should tear a moon apart. They may have been spared because of their small size. The stay of execution is temporary, however. Metis and Adrastea are so near to Jupiter that they circle the planet faster than Jupiter rotates. In time, their orbits will decay, and the pair will tumble to their doom into Jupiter's atmosphere.

These four moons, collectively known as the Amalthea Group, are immersed in Jupiter's faint ring system. Astronomers theorize that tiny particles blasted off these moons by impacts with micrometeorites contribute material to the rings.

THE GALILEAN SATELLITES: JUPITER'S "BIG FOUR"

Jupiter's next satellite group deserves the nickname the "Big Four." Large enough to be easily detected by Galileo's crude telescope in 1610, Io, Europa, Ganymede, and Callisto are giants as moons go. One of them, Ganymede, is larger than the planet Mercury. The remaining three rank among the solar system's six largest moons. The Galilean satellites have orbits that are essentially lined up with Jupiter's equator, evidence that they formed together with Jupiter. To understand and appreciate the true nature of these worlds, let's examine each through the eyes of the *Voyagers* and the *Galileo Orbiter* that followed.

Io, the Volcanic Moon

Pronounced: EYE-oh
Diameter: 3,630 km (2,250 mi)
Mean Distance from Jupiter: 421,600 km (261,400 mi)
Orbital Period: 1.8 days
Period of Rotation: 1.8 days (synchronous)
Average Density: 3.5 g/cm^3
Surface Gravity, Compared to Earth: 18.4%

The book *Cosmos* by Carl Sagan contains a chapter called "Heaven and Hell." In it, Sagan describes the hellish environment found on the surface of Venus. The planet's searing-hot surface temperatures, crushing atmospheric

Figure 5.2 Io. AP Photo/NASA Jet Propulsion Laboratory.

pressure, and sulfuric acid mist combine to create a picture right out of the pages of Dante's *Inferno*. However, there is another solar system body that might rival Venus for the title of most hellish location—Jupiter's moon Io.

Io is an unimaginably hostile world, where a combination of intense radiation and high-voltage electrical current would prove instantly fatal to an astronaut standing on its surface. Even if he or she were shielded from these hazards, the landscape itself would prove to be anything but heavenly. Huge sulfur-belching volcanoes cover Io's surface with this foul-smelling substance. Io's atmosphere, thin as it is, consists of noxious sulfur dioxide. Just standing on Io would be a test of an astronaut's courage. Stress on Io's crust, caused by Jupiter's gravity, would cause the ground to constantly shake underfoot. Since Io's surface gravity is about one-sixth Earth's, the much-lighter astronaut would have trouble maintaining his or her footing.

How is it possible for such hellish conditions to exist in the cold depths of the solar system? After all, much of Io's surface is locked in a deep freeze, with temperatures of -150°C (-238°F). The answer lies not with the too-distant Sun, but with Io's proximity to Jupiter and with the gravitational pull of two other Galilean moons.

The closest of the Galilean moons to Jupiter, Io is deeply immersed in Jupiter's powerful magnetosphere. As Jupiter rotates, the magnetic field lines sweep across Io, stripping the moon of about 1,000 kg (over one ton) of its matter every second. This material, mainly sulfur and oxygen, becomes ionized by the magnetic field and forms a highly radioactive plasma torus centered on Io's orbit. The magnetic field lines also generate a strong electrical current that follows the magnetic field lines directly from Io to Jupiter's atmosphere.

The root of Io's volcanism lies in its orbit. Io travels around Jupiter in a synchronous orbit with a period of 1.8 days. The side facing Jupiter is constantly pulled by the big planet's gravity. The situation is compounded by the fact that Jupiter's next two moons outward, Europa and Ganymede, revolve in what is known as a **Laplace, or orbital, resonance** with Io. Europa revolves around Jupiter once every 3.6 days—exactly twice Io's period of revolution. Ganymede's 7.2-day orbital period is four times Io's. Each time Io orbits Jupiter twice, it finds itself in a gravitational tug-of-war between Jupiter and Europa. Four orbits bring Io in line with Ganymede. These periodic alignments cause a **tidal flexing** of Io's crust, lifting it up to 100 meters (330 feet) high—the height of a 33-story skyscraper! The tidal flexing generates heat in Io's interior much in the way a wire coat hanger gets hot when it is repeatedly bent back and forth.

It was the images of Io relayed back to Earth by *Voyagers 1* and *2* in 1979 that ended forever the idea that planetary moons are uninteresting places. Instead of our Moon's static, cratered terrain, overlain with drab shades of gray, *Voyager* photographs revealed a landscape rich in color—a veritable palette of white, yellow orange, and red, interspersed with black pockmarks.

Its appearance was variously described as that of a cheese pizza covered with slices of black olive or a moldy orange. One *Voyager* team scientist quipped that we might not know what was wrong with Io, but a dose of penicillin would help!

There was something else striking and unexpected about the *Voyager* images. Io's terrain had few, if any, craters. All bodies in the solar system underwent an early period of meteoric bombardment and cratering. A lack of impact craters indicates a young surface, where subsurface material of some form resurfaced the landscape. Why is Io's surface so geologically young? What substances cause Io's unusual colors?

The answers came when an enhancement of one of the *Voyager* images revealed a gigantic volcanic plume rising 300 kilometers (185 miles) above Io's surface. This is a dozen times higher than the volcanic plumes generated by Earth's volcanoes. Closer inspection of the *Voyager* photos revealed Io to be the most volcanically active body in the solar system—a world whose surface is littered with active volcanoes and fissures. Spectral analysis of Io revealed that the colors are consistent with sulfur and sulfur compounds at various temperatures from molten to solid.

The *Voyager* flybys uncovered about a dozen active volcanoes on Io. The *Galileo Orbiter*, which arrived at Jupiter 16 years later, revealed more than a hundred. Among the most prominent of Io's volcanoes are Pele, Loki, and Prometheus. Temperature readings on Io's surface taken by the *Galileo Orbiter* revealed that the lava escaping some of its vents was too hot to be made of sulfur. This material is more likely silicate-based, like the lava associated with Earth's volcanoes.

Not all of Io's mountain peaks are volcanic in origin. The highest, Boosaule Mons, towers 16,700 meters (55,000 feet) above the surrounding terrain. By comparison, Earth's loftiest peak, Mt. Everest, is about half that high. While Mt. Everest was formed by the movement of plates in the Earth's crust, Boosaule Mons may have been created when a pileup of sulfur deposits in one area caused uplift in Io's crust elsewhere.

Mapping Io's ever-evolving surface is an ongoing task. Dramatic changes were evident when photos taken with the *Galileo* probe in 1999 were compared with those made by the *Voyagers* a mere two decades earlier. Astronomers envision a fascinating scenario, where Io's surface rapidly accumulates material ejected from its volcanoes. Each time a volcano erupts, the original ejected material gets buried deeper and deeper. Ultimately, it reaches a lower level where internal heat melts it, and it finds its way to another volcano, where it is erupted back to Io's surface. Io accomplishes the ultimate act of recycling, literally turning itself inside out!

Io is slightly larger and more massive than our Moon. It differs from many of the ice-laden satellites in the outer solar system, having instead a rocky "terrestrial planet-like" structure. Its average density, a strong indicator of its overall composition, is similar to our own Moon, slightly greater than Europa, and roughly twice that of Ganymede and Callisto.

Galileo Orbiter measurements of Io's gravity and the detection of a possible magnetic field allowed astronomers to generate a model of its interior. At the center is a metallic nickel-iron core that extends halfway to the surface. The core is surrounded by a rocky silicate shell, molten on the interior and solid at the surface. This layered structure is a result of a process called **differentiation**. Differentiation is possible only if a body cools slowly enough to allow dense materials to settle in its core.

Why does Io lack the water ices found in such abundance on its kindred Galilean moons? One explanation is that when the solar system was still in its formative processes and these moons were forming around Jupiter, heat from a young Jupiter drove **volatiles** (light materials like water) from Io. Left behind was heavier, rocky, and metallic material. Further away from Jupiter's heat, the outer moons were able to retain these ices. The analogy is similar to the formation of the solar system, where the Sun's radiant energy drove light materials from inner regions, leaving behind dense, rocky inner (terrestrial) planets and lighter (Jovian) outer planets. Any ices that did remain with Io were ultimately belched out of its volcanoes and dissipated into space.

It may seem that poor Io gets bullied around by Jupiter, which warps the moon's crust and steals tons of its material, but the moon does get some measure of payback. Some of the ionized particles in the Io torus get carried by Jupiter's magnetic field lines to the planet's surface. They bombard Jupiter's upper atmosphere, creating auroras. Meanwhile, the electric current that flows along magnetic field lines from Io to Jupiter creates powerful lightning bolts in Jupiter's atmosphere.

A Day on Io

Good morning, and welcome to Io! You're standing on the side of Io that faces Jupiter. Thanks to late twenty-first-century technology, you're fully protected from the intense radiation that bathes this moon's surface. Right now, Jupiter looms high above at half-moon phase, its sunlit side facing the direction where the Sun will rise. Although Jupiter is as far away as the Moon is from Earth, the huge planet appears 40 times larger than the full Moon. Your fingers extended at arm's length fail to cover its disk. For a moment, you face eastward and spot a bright bluish "star" above the horizon. You feel a touch of homesickness. Earth seems so far away! In a few minutes, the rising Sun appears as a blinding flash of light. You're five times farther from this fiery furnace than you were on Earth, and the Sun's disk seems proportionally smaller. But it's still a blindingly bright object, and it's wise to avoid staring directly at it.

As the morning wears on, the Sun rises higher in Io's sky. Meanwhile, Jupiter assumes the shape of an increasingly thin crescent. Near midday, the Sun seems to touch the side of Jupiter—the beginning of a two-and-a-half hour eclipse. A ring of brilliant red surrounds the planet—sunlight refracted by Jupiter's atmosphere. With Jupiter's nighttime side facing us, we see ghostly auroral displays near the poles. Eerily beautiful are the multitude sparks of light from thunderbolts generated in Jovian storms. They blink like a myriad of fireflies on a midsummer evening. Looking away from Jupiter, we see the stars in Io's airless sky. A voyage of hundreds of millions of kilometers from Earth hasn't altered the configurations of the constellations. The Big Dipper, Polaris, and Cassiopeia remind you of the evenings you spent under the stars back on Earth. Eventually, the Sun reappears on the other side of Jupiter, and Io's landscape is again bathed in sunlight.

Throughout the afternoon as the Sun sinks lower in the sky, Jupiter's sunlit portion again grows. By sunset, Jupiter is again a half moon, this time facing the western horizon as if to catch a last glimpse of the Sun. Nighttime on this part of Io will never bring the kind of darkness we experience during Earthly evenings. The later in the evening, the brighter Jupiter gets as it approaches full-moon phase. By the middle of the night, Jupiter is an incredible sight, its banded atmosphere laid out before you in a riot of colors and writhing cloud patterns. A black dot slowly moving across the Jovian disk marks Io's shadow.

You could remain up all night, but over 30 hours have passed since sunrise. A day on Io is almost twice the length of an Earth day. You head back to your room on Io Base, so exhausted that even the constant ground tremors fail to interrupt your sleep.

Europa, the Ocean Moon

Pronounced: "you-ROH-puh"
Diameter: 3,130 km (1,940 mi)
Mean Distance from Jupiter: 670,900 km (416,900 mi)
Orbital Period: 3.6 days
Period of Rotation: 3.6 days (synchronous)
Average Density: 3.0 g/cm^3
Surface Gravity, Compared to Earth: 13.5%

Figure 5.3 Europa. AP Photo/NASA Jet Propulsion Laboratory.

To move from Io to Jupiter's next major moon, Europa, is the equivalent of putting down a bizarre science fiction book and picking up an intriguing mystery novel. The smallest of the Galilean satellites with a diameter a bit less than our Moon's, Europa imitates Io in several ways. It's relatively dense (indicating a metallic/rocky interior), its surface is young and sparsely cratered, and it suffers tidal flexing (this time from Jupiter and Io on one side and Ganymede on the other). Europa also has a synchronous orbit around Jupiter, and that orbit is immersed in Jupiter's radiation belt.

Compared to Io's multicolored pizza pie exterior, Europa appears rather bland and featureless. Its surface is extremely bright, one of the most reflective in the solar system. While Io's surface is littered with towering volcanoes that layer its surface in silicates and sulfur compounds, Europa presents an amazingly flat, ice-covered surface riddled with cracks hundreds of kilometers long. Some of the cracks are tainted with a light reddish-brown material. Topographic features on Europa rarely rise higher than a few hundred meters.

Europa has an extremely thin atmosphere comprised almost totally of oxygen. The oxygen in our atmosphere has a biological origin, generated by plants during photosynthesis. You won't find lush green forests on Europa, however—not in an environment where temperatures never exceed $-140°C$ ($-225°F$). Europa's atmospheric oxygen has a decidedly nonbiologic source. Electrically charged particles from Jupiter's radiation belt bombard Europa's icy surface, releasing water vapor, which separates into hydrogen and oxygen. The hydrogen escapes into space, leaving behind a trace of oxygen.

Europa's internal structure was indirectly investigated in a similar manner to the way Io's interior was surveyed—by analysis of its gravitational and magnetic fields by the *Galileo* probe. Recent models based on the *Galileo* data show that Europa very likely has a small metallic core overlain by rock and soft ice, surrounded by a zone of salty water and topped by a crust of impure water ice. This layered structure is again a result of differentiation.

How deep is Europa's icy crust? Its thickness is hinted at by the nature of what few meteoric craters cover Europa's surface. An extremely thin crust would allow meteors to plummet through, leaving behind a hole that would have filled in with subsurface water. If the crust were extremely thick, impact craters would be several kilometers deep, like the ones on our moon. What few craters are found on Europa's surface seem to fit an in-between scheme. Another indicator of crust thickness lies in the appearance of Europa's broken surface. It's strikingly similar to Earth's Arctic regions, where ice sheets a few kilometers thick break apart and reassemble. Based on these forms of evidence, astronomers surmise that Europa is cloaked in a layer of water ice anywhere from 5 to 20 kilometers (3 to 12.5 miles) thick.

As for what lies directly beneath the crust, astronomers speculate that there may be an ocean of liquid or slushy salt water as much as 100 kilometers (62 miles) deep. This idea is based in part on the presence of an extremely weak magnetic field—possible only if there were some kind of salty liquid water to generate electric current and a magnetic field. If this scenario proves to be true, Europa could hold more than two times as much water as our Earth.

Astronomers can only guess at the chemical nature of Europa's subsurface ocean. Clues may be found with mysterious-looking brownish areas, including the cracks that crisscross Europa, first imaged by the *Voyagers* in 1979. Higher resolution photos from the *Galileo Orbiter* added clusters of 10-kilometer (6 mile)-wide "freckles," or **lenticulae**, which also have a brownish hue. Several large areas of brownish-colored terrain, including Thera and Thrace, are also present. Thera is slightly larger than Rhode Island, and Thrace is about twice as large. All of these features seem to have been created when warm subsurface water broke through to the surface and quickly froze, leaving behind traces of minerals—possibly sulfur compounds.

How is it possible for water to exist in a liquid state so far from the Sun? The answer lies in the same mechanism that generates Io's internal heat—tidal forces from Jupiter and a gravitational tug-of-war from nearby moons—this time Io and Ganymede. Europa's orbital period of 3.6 days is exactly twice Io's and half Ganymede's. These competing pulls flex Europa, creating an internal heat that allows the subsurface water to remain liquid.

It is within this subterranean sea that Europa takes on the element of a mystery novel. Being one of the few bodies in our solar system besides Earth that contains an abundance of liquid water, Europa is a potential abode for extraterrestrial life. Near the end of the *Galileo Orbiter*'s Jupiter mission, this possibility forced NASA scientists to program the probe's path so it would fly into Jupiter and burn up in the planet's atmosphere. No one wanted to risk the possibility that *Galileo* might crash-land on Europa, contaminating it with any Earthly organisms that might have hitched a ride.

For centuries biologists believed that life as we know it could only exist where there was a combination of liquid water, a nutrient source, organic material, and (most importantly) sunlight. Sunlight was needed, they said, to generate photosynthesis in plants that would form the base of a food chain. The belief was made moot by the discovery of colonies of deep-sea marine life existing near geothermal vents in the pitch-black abyss of our oceans. Chemicals pouring out of the vents support bacteria that replace plants as the food base for this undersea life. Is it possible that similar vents, triggered by the same tidal forces that made Io volcanic, exist at the bottom of Europa's sea? Could these in turn have promoted the birth of life in this Jovian moon? As of yet, we have no way of answering these questions. Until we further explore Europa, they will remain tantalizing questions.

Ganymede, King of the Moons

Pronounced: "GAN-uh-meed"
Diameter: 5,260 km (3,280 mi)

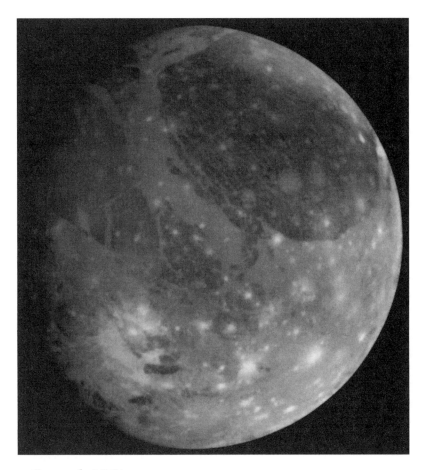

Figure 5.4 Ganymede. NASA.

Mean Distance from Jupiter: 1,070,000 km (663,400 mi)
Orbital Period: 7.2 days
Period of Rotation: 7.2 days (synchronous)
Average Density: 1.9 g/cm^3
Surface Gravity, Compared to Earth: 14.6%

How appropriate that the King of the Planets be attended by the King of the Moons. This Goliath is larger than the planet Mercury! Indeed, Ganymede would be classified as a planet if it orbited the Sun instead of Jupiter. A line the length of our Moon's diameter, traced out on a map of the United States, would stretch from Boston two-thirds of the way across the country to the Rocky Mountains. Were that line equal to Ganymede's diameter, it would cross the entire United States from coast to coast.

Before the *Voyager* missions gave us close-up views of Ganymede, astronomers were intrigued by its mottled appearance when viewed with large earthbound telescopes. It wasn't an illusion. Photos taken by the *Voyagers* revealed a moon whose surface is a blend of dark and light areas.

In some ways, Ganymede bears a strong resemblance to our Moon, whose light-hued highlands are interrupted by the dark lunar seas. But, there is a distinct difference. On our Moon, the light areas represent ancient cratered highlands, while the dark seas are great basaltic lava plains that resurfaced much of the Moon much later. On Ganymede, the dark regions, called **regios**, represent the most ancient terrain. We know this because the regios are highly cratered. Their dark appearance results from a coating of impact-generated dust. Bright circular areas called **palimpsests** appear to be impact craters that filled in with subsurface material.

Ganymede's lighter, younger terrain is punctuated by long ridges and grooves that may have been created by tectonic movement. Overall, Ganymede's surface is far more ancient and heavily cratered than either Io's or Europa's.

In 1996, instruments on board the *Galileo Orbiter* detected a dipolar magnetic field surrounding Ganymede. It's the only one found among the moons in the solar system. Would future astronauts be able to use Ganymede's magnetic field to navigate around its surface with compasses? Probably not. Ganymede and its magnetic field are immersed in Jupiter's magnetosphere.

Ganymede is the outermost of the Galilean satellites to orbit within the confines of Jupiter's radiation belt. Like Europa, Ganymede has an extremely tenuous atmosphere of oxygen, created by bombardment of its water ice surface by ionized particles in Jupiter's magnetosphere.

Ganymede has something else not detected on any other Jovian moons—aurorae. They were discovered by a team of astronomers using the Hubble Space Telescope. Charged particles from Jupiter's magnetosphere are captured by Ganymede's magnetic field and drawn toward the moon's north and south poles. The particles interact with Ganymede's thin oxygen atmosphere to produce extremely faint auroral displays. Unlike Earth's northern and southern lights, Ganymede's aurorae are too faint to be seen visually.

....

Moon or Planet?

It's logical to ask why Ganymede, which is larger than Mercury, is classified as a moon and not a planet. Unfortunately for the giant moon, being big won't gain it admittance into the exclusive Planet Club. One of the factors that determine planetary status is orbit. If a body circles the Sun (and has enough size), it's considered a planet. If it revolves around a planet, it's a moon regardless of its size. So Ganymede, which orbits Jupiter instead of the Sun, is a moon. Case closed.

Here's an eye-opening thought. Were our Earth placed near Jupiter (an unpleasant situation, as the big planet's gravity would create massive earthquakes and raise the ocean tides' high enough to drown coastal cities, while the radiation belt would "cook" all life on Earth), we would begin orbiting the big planet. Earth would become another of Jupiter's swarm of moons.

....

With an average density only two-thirds that of Io and Europa, Ganymede is likely a 50-50 mix of rock and ice. Models based on data relayed by the *Galileo* probe suggest a molten iron-rich core, a lower mantle of rocky silicates, an upper mantle of ice, and a water ice crust. There is evidence that Ganymede, like Europa, has a salty subsurface ocean.

Callisto, a Bombarded Moon

Pronounced: "kah-LISS-toe"
Diameter 4,800: km (2,975 mi)
Mean Distance from Jupiter: 1,883,000 km (1,167,500 mi)
Orbital Period: 16.7 days
Period of Rotation: 16.7 days (synchronous)
Average Density: 1.9 g/cm^3
Surface Gravity, Compared to Earth: 12.6%

Figure 5.5 Callisto. AP Photo/NASA Jet Propulsion Laboratory/DLR.

Callisto, outermost of Jupiter's Galilean satellites, is the third largest moon in the solar system (behind Ganymede and Saturn's moon Titan). It's slightly smaller than the planet Mercury, but only one-third as massive. Pictures from the *Voyager* and *Galileo* missions reveal a relatively dark surface pockmarked by bright craters. Callisto is, in fact, the most heavily cratered body in the solar system.

Callisto's craters are similar in appearance to the bright-rayed craters visible on our Moon around full phase. Rayed craters result when an impacting object punctures the dark surface, ejecting brighter subsurface material.

Rayed craters aside, our Moon and Callisto are hardly twins. Compared to our silicate-laden Moon, icy Callisto is only half as dense. Callisto lacks evidence of geologic upheaval. There are no large basalt-based lava plains like the ones that created our Moon's dark seas. Mountain ranges like the lunar Alps and Apennines are also absent. While our Moon and Callisto likely formed at the same time, ours underwent major physical changes a billion years after its birth, while Callisto appears to have been relatively unchanged for the past 4 billion years. It's fair to say that this bombarded moon is a "fossil" whose surface is frozen in time.

Any change in Callisto's surface is a result of meteoric impact or the slow **sublimation** of its surface ice. Close-up photos of Callisto, obtained by the *Galileo Orbiter,* show a surprising lack of craters smaller than one kilometer (one-half mile). There is an abundance of bumps and knobs 80 to 100 meters (260 to 325 feet) high. They may be the remains of ancient crater walls whose material has slowly eroded by sublimation. The topography of many of Callisto's largest craters is likewise "softened," and palimpsests like the ones that exist on Ganymede have been found on Callisto.

Among Callisto's more prominent surface features is a pair of huge **multi-ring basins** similar to our Moon's Mare Orientale. Each is the site of an impact by a huge cosmic body that created a series of bull's-eye ridges around a central basin. The larger of the two is Valhalla, a series of concentric rings over 3,000 kilometers (1,800 miles) across. The other, Asgard, is half the size of Valhalla, but still big enough to encompass the entire state of Alaska.

Callisto is also noted for its interesting crater chains. About a dozen have been imaged by the *Voyagers* and the *Galileo Orbiter.* The largest is the Gipul Patera, an eye-catching chain that cuts 620 kilometers (385 miles) across Callisto's surface. On Earth, the Gipul Patera would run along the United States' East Coast from Boston, Massachusetts, through New York, Philadelphia, and Baltimore to Washington, DC. Callisto's crater chains were quite likely created by impacts from fractured comets similar to the Shoemaker-Levy 9 Comet that struck Jupiter in 1994.

Data gleaned by the *Galileo Orbiter* indicate that Callisto is covered by an icy crust 200 kilometers (124 miles) thick. The rest of this moon's interior is a mix of ice and rock, with the concentration of rock increasing toward Callisto's center. There is evidence for a 10-kilometer (6-mile)-deep layer of salty water between the crust and ice/rock interior.

The *Galileo Orbiter* also discovered a very tenuous atmosphere of carbon dioxide surrounding Callisto. The moon's weak gravity would not be able to hold these gases for more than a few days, so there must be a source to replenish it. The same sublimation process that has slowly eroded Callisto's surface may be converting frozen carbon dioxide into its gaseous state.

Because its orbital period of 16.7 days isn't in resonance with Io, Europa, and Ganymede, Callisto doesn't suffer from tidal disruptions like the other three. It's also the only one of the Galilean satellites that doesn't lie in Jupiter's radiation belt. For these reasons, Callisto may become the home base for a future manned expedition to Jupiter.

Jupiter's Moons and the Speed of Light

There is more to the historical importance of the Galilean moons than proof that celestial bodies can orbit something besides the Earth. These moons also helped scientists make the first reasonable calculation of the speed of light.

In the decades after Galileo had discovered Jupiter's four bright moons, astronomers followed their motions around the big planet. Because their orbits are edge-on to our line-of-sight, the moons appear to dance back and forth from one side of Jupiter to the other, alternately passing in front of and behind the planet. The Italian astronomer Giovanni Cassini began making detailed observations and timings of those occasions when Jupiter's closest Galilean satellite, Io, was eclipsed (passed behind) by Jupiter. He began to notice that discrepancies in the predicted times of these eclipses grew larger whenever Jupiter drew near to the Earth or drifted farther away. Cassini reasoned that light must travel at a finite speed, and that as the distance between Jupiter and the Earth changed, so did the amount of time it took light from the eclipse to reach our eyes.

In 1671, Cassini moved to the Paris Observatory, and was joined the next year by Danish astronomer Ole Roemer. Roemer made more observations of eclipses of Jupiter's moons and noticed that they occurred sooner than predicted as the Earth approached Jupiter and later as the two planets parted. Although Roemer never made an estimate of the actual speed of light, astronomers in the early 1800s repeated the work of Cassini and Roemer and deduced that light must travel at a speed of around 300,000 kilometers (186,000 miles) per second—close to today's accepted value.

THEMISTO, A LOST AND FOUND MOON

Themisto (theh-MISS-toe) might be considered one of Jupiter's loneliest satellites. While most of the moons in Jupiter's impressive retinue belong to orbital groups, Themisto travels alone in the middle of a 9,300,000-kilometer (5,780,000-mile) gap separating the Galilean moons and the Himalia Group.

Themisto played a cosmic version of hide-and-seek with astronomers. It was originally discovered by Charles Kowal and Elizabeth Roemer on a photographic plate taken at Mount Palomar in September 1975, and given the provisional designation S/1975 J 1. Unfortunately, not enough observations were made to allow for the determination of an accurate orbit. The tiny moon was soon lost and remained so until 2000, when a team of astronomers led by Scott Sheppard and David Jewitt of the University of Hawaii recaptured it.

Little is known about Themisto other than the fact that it has a diameter of about 8 kilometers (5 miles) and orbits Jupiter once every 130 days at an average distance of 7,507,000 kilometers (4,665,000 miles). Its orbit is inclined nearly 46° to Jupiter's equator.

Beyond Themisto is a swarm of tiny moons, many with diameters less than 8 kilometers (5 miles). Most of these moons were found by a team of astronomers from the University of Hawaii, led by Scott S. Sheppard. From 2000 to 2003, they used wide-field CCD imagers coupled to large reflecting telescopes on Mauna Kea, Hawaii, to track down over 40 new moons. Their extremely dark surfaces and highly **inclined orbits** indicate that most, if not all, may be captured bodies.

···

Scott S. Sheppard (1976–)

Which astronomer holds the record for discovering the most solar system satellites? Not Galileo! He was the first, capturing four bright moons around Jupiter in 1610, but he didn't find the most. That honor goes to astronomer Scott S. Sheppard. Dr. Sheppard grew up in the wide-open spaces of Nebraska, where beautifully dark starlit nights captured his imagination at an early age. In 1998, he received an undergraduate degree in physics from Oberlin College. Six years later, he completed his graduate studies in astronomy at the University of Hawaii.

While at the University of Hawaii, Sheppard led a team of astronomers in a search for moons around the outer planets. Using wide-field CCD cameras coupled with large reflecting telescopes on Mauna Kea, Hawaii, his group corralled dozens of tiny, extremely faint satellites in orbit around Jupiter, Saturn, Uranus, and Neptune. He believes that each of the giant planets will ultimately be found to harbor about one hundred moons larger than one kilometer (one-half mile) in diameter.

···

THE HIMALIA GROUP

The Himalia (hih-MAL-yuh) Group consists of five Jovian moons, represented by Himalia, and including Leda (LEE-duh), Lysithea (ly-SITH-eel-uh), Elara (Eel-lar-uh), and the as-yet unnamed S/2000 J 11. They fill the zone 11,170,000 to 12,560,000 kilometers (6,940,000 to 7,805,000 miles) from Jupiter, with orbits inclined an average of 27.5° from Jupiter's equator.

Himalia itself was discovered in 1904 photographically by Charles Dillon Perrine at the Lick Observatory. Sixth largest of Jupiter's satellites, this irregularly shaped moon is 170 kilometers (106 miles) along its greatest dimension. It orbits Jupiter once every 250.6 days. Elara (discovered by Perrine in 1905), Lysithea (Seth Nicholson, 1938), and Leda (Charles Kowal, 1975) have diameters between 10 and 86 kilometers (6 to 53 miles), while S/2000 J 11 (Scott Sheppard/University of Hawaii team, 2000) is perhaps 4 kilometers (2.5 miles) across.

CARPO

Carpo (CAR-po) is the second of Jupiter's orbitally independent moons; the other one is Themisto. Carpo was discovered in 2003 by the same Scott Sheppard-led University of Hawaii team whose survey had recovered Themisto three years earlier.

Orbiting Jupiter at an average distance of 16,990,000 kilometers (10,577,000 miles), this tiny 3-kilometer (2-mile)-wide moon takes 15 months to circle the big planet. Carpo's elliptical, highly inclined orbit is unstable and may lead to its ultimate demise. The tiny moon may one day stray within range of the Galilean satellites, where it will either collide with one of them or be gravitationally flung out of the Jovian system.

THE ANANKE GROUP

From here on out, all of Jupiter's moons revolve in a backward, or retrograde, direction and in highly inclined orbits—strong evidence that they are bodies from other parts of the solar system that strayed too close to Jupiter and became prisoners of its gravity. Most of these moons were discovered between 2000 and 2003 by a pair of teams using sensitive CCD cameras coupled to huge telescopes on Mauna Kea, Hawaii. One team was directed by Scott Sheppard and David Jewitt, of the University of Hawaii; the other team was led by Brett Gladman, of the University of British Columbia, Canada.

A 1.9 million kilometer (1.2 million mile)-wide zone, beginning a little over 19,000,000 kilometers (11,800,000 miles) from Jupiter is home to the Ananke (a-NANG-kee) Group of Jovian moons. It consists of some 16 positive and suspected members, highlighted by the 28 kilometer (17.4 mile)-wide Ananke. Discovered in 1951 by Seth Nicholson, at the Mount Wilson Observatory in California, Ananke swings around Jupiter once every 21 months. As small as Ananke is, it's a veritable giant when compared to the remaining moons in its group. The rest have diameters ranging from 2 to 7 kilometers (1.2 to 4.3 miles). All members of the Ananke Group have orbits with inclinations around 150° to Jupiter's equator.

THE CARME GROUP

Seventeen moons are currently recognized as definite or suspected members of the Carme (KAR-mee) Group. Spread out over a zone 1,112,000 kilometers (690,000 miles) wide, beginning 22,931,000 kilometers (14,250,000 miles) from Jupiter, this group mirrors the Ananke Group in that it is dominated by one moon (in this case Carme) that is substantially larger than the rest. They also have retrograde orbits, but the inclinations are slightly greater than the Ananke Group, typically 165°.

Carme was discovered by Nicholson at Mount Wilson in 1938. None of the remaining moons in the Carme group has a diameter greater then 5 kilometers (3 miles). Carme itself is 46 kilometers (28.6 miles) in diameter and orbits Jupiter in a retrograde direction once every two years.

THE PASIPHAE GROUP

Pasiphae (pah-SIF-ah-ee) was first seen on a photographic plate taken at the Royal Greenwich Observatory by Philibert Jacques Melotte in 1908. The 60-kilometer (37-mile)-wide moon gives its name to a group that potentially contains 14 known Jovian moons orbiting between 21,263,000 and 24,543,000 kilometers (13,212,000 to 15,250,000 miles) from Jupiter. All have retrograde orbits with inclinations between 140° and 160°. Pasiphae is so far from Jupiter that it takes about two years to complete an orbit.

Among the moons in the Pasiphae Group is the 38-kilometer (24-mile)-wide Sinope (sah-NOH-pee). Discovered on a photographic plate taken by Seth Nicholson in 1914, it was for many years thought to be Jupiter's outermost satellite until five tiny, more distant moons were swept up between 2001 and 2003 by the Scott Sheppard team. Jupiter's currently recognized most-distant moon (in fact, the remotest moon in the solar system) is S/2003 J 2. This 2-kilometer (1.2-mile) chunk of rock and ice is 28,570,000 kilometers (17,753,000 miles) from Jupiter—a whopping 80 times greater than the Earth-Moon distance. An orbital swing around Jupiter takes nearly three years.

JUPITER'S MOONS: THE MOST POPULAR FUTURE TOURIST SITES

A vacation spent viewing one of the volcanoes on Io may seem a bit improbable to those of us living in the early twenty-first century. However, the idea of visiting Hawaii's Kilauea Volcano would have seemed equally far-fetched to someone living in Renaissance Europe at the time Galileo was discovering Jupiter's moons. Assuming that future technology will make space travel quick and safe, here are some excerpts from twenty-second century travel brochures.

1. Amalthea. If you think standing on the rim of the Grand Canyon offers a magnificent visual panorama, you haven't viewed Jupiter from one of its nearest moons, Amalthea. Few sights in the solar system are more exciting and dramatic than this giant planet as it looms large and majestic in the Amalthean sky. A 12-hour stay on Amalthea allows you time to circle Jupiter one full time. For a change of pace, you can play a game of "Amalthea-Hop."

Because of this moon's low-gravity environment (the average adult weighs less than a newborn kitten) you can cover several hundred meters in a single jump.

2. Io—Boosaule Mons. Enjoy mountain climbing? Mt. Everest is nothing compared to Io's tallest peak, Boosaule Mons. Its 16,700-meter (55,000-feet) elevation is nearly double Everest's height. If the task of climbing Boosaule seems too daunting, remember—on Io, you only weigh one-fifth what you weigh on Earth. Hike up Boosaule Mons and enjoy a view that is literally out of this world!

3. Io—Tvashtar Paterae. Think the Hawaiian volcanoes like Kilauea are hot? Try Io's Tvashtar Paterae. Named after the Hindu god of blacksmiths, Tvashtar has it all—a 25-kilometer (15.5-mile)-long lava-spewing vent and a huge lake of superhot silicate lava for starters. And when Tvashtar blows its top—watch out! Lava erupts in a spectacular fountain that arches nearly 400 kilometers (250 miles) in the air, and covers an area the size of Texas. Visit Tvashtar now, while it's still hot!

4. Europa—Thera. Interested in some serious oceanographic research? Visit Europa's Thera Base, the Wood's Hole of outer space. Thera is an area of Europa's icy crust the size of Rhode Island that sank because of an upwelling of warm water from below. The thin ice here allows scientists easy access to Europa's deep subsurface sea. Join a research team as it analyzes the chemistry of Europa's water. Take a one-hour trip in a submersible as it seeks European life forms. Experience the excitement of discovery!

5. Callisto—the Ridges at Valhalla Basin. As a kid, did you enjoy roller coaster rides at the local amusement park? Take the thrill of a lifetime as you speed over the ridges of Callisto's famous Valhalla Basin. Billions of years ago, a huge cosmic body slammed into Callisto, creating the moon's signature Valhalla Basin—a series of concentric "bull's-eye" ridges 3,000 kilometers (1,860 miles) across. Starting from the center of the basin, we'll speed outward by rocket sled on the Valhalla Ridge Road. The ride from one ridge to the next will make the scariest roller coaster seem tame by comparison, and the view at the last ridge will leave you breathless.

6. Callisto—Gipul Catena. Remember that string of comets (Shoemaker-Levy 9) that crashed into Jupiter way back in 1994? The impact areas were quickly absorbed by Jupiter's dense atmosphere, leaving nary a trace. What if that fractured comet had struck a solid surface? Guess no more—Callisto's fabulous Gipul Catena has the answer. If you thought Arizona's Meteor

Crater was something, try dozens of even bigger craters, all lined up in a remarkable chain that runs for 620 kilometers (385 miles) across Callisto's frozen surface. If you're a distance runner, why not enter the Gipul Marathon, a 42-kilometer (26.2-mile) run across the largest crater in the Gipul Catena? In case our Gipul Tour is completely booked, we offer alternate trips to several other crater chains on Callisto.

WEB SITES

http://www.nineplanets.org/jupiter.html.
    Bill Arnett's Nine (8) Planets Web site contains information on the planet's moons.
http://solarsystem.nasa.gov/planets/profile.cfm?Display=Moons&Object=Jupiter.
    A look at Jupiter's moons from a NASA perspective.
http://www.dtm.ciw.edu/sheppard/satellites/jupsatdata.html.
    Regularly updated data on Jupiter's moons, prepared by astronomer Scott Sheppard.
http://www.astronomycast.com/astronomy/episode-57-jupiters-moons.
    The written script of a podcast on Jupiter's moons, done by astronomers Fraser Cain and Pamela Gay. Contains links to other Jupiter satellite Web sites, as well as a link to the actual podcast.

6

Saturn, Crown Jewel of the Solar System

August 25, 1981: Two years after its Jupiter encounter, *Voyager 2* has arrived at Saturn. Cruising 41,000 kilometers (26,000 miles) above Saturn's cloud tops, the craft is witness to a panoramic view of the Ringed Planet. Spectacular images of Saturn, its rings, and moons will augment those taken by *Voyager 1* a few months earlier. These are exciting times for planetary scientists.

Saturn is eight-year-old Michelle's favorite telescopic sight. She knows that in third grade she'll have to write a report on one of the planets. Saturn will be her choice. Pictures of the Ringed Planet and some of its moons will appear in the popular astronomy magazines in a few months. Michelle will use them to supplement her report. This young girl will know more about Saturn than great past astronomers like Galileo.

THE RINGED PLANET

How knowledgeable is the average man or woman on the street about astronomy? Probably not very. Astronomy is a subject about which much of the populace professes a great deal of ignorance. Relatively few people can explain what a black hole is, let alone name the planets in our solar system in correct order. Nevertheless, we can stop virtually anyone on the street, ask him or her to describe the planet Saturn, and almost certainly get the quick response, "It's the planet with rings."

Saturn Data

Period of Revolution	29.46 years
Period of Rotation	$10^h\ 39^m$
Axis Tilt	26.7°
Equatorial Diameter	120,536 km (74,732 mi)
Mass (Earth = 1)	95.2
Surface Gravity* (Earth = 1)	1.2
Density (water = 1.0 g/cm^3)	0.7
Number of Moons	60
Mean Distance from Sun	9.5 AU (1,429,400,000 km [857,640,000 mi])

* Since none of the Jovian planets has a solid surface, gravity is calculated at the visible cloud tops.

The fact that Saturn is a planet encircled by rings is a piece of astronomical knowledge most of us learn in grade school and then have reinforced throughout our educational years. By adulthood, we've been programmed to relate the words *Saturn* and *rings* just as surely as we connect the words *Moon* and *craters*. The reason becomes obvious whenever you look at a photo of Saturn in any astronomy textbook or magazine. The Ringed Planet is a wondrous sight that leaves an indelible image on our minds.

Saturn's rings capture our attention and imagination so completely that we scarcely pay attention to the planet itself. It's that pale-yellow, slightly flattened ball nestled inside the rings. Amateur astronomers are easily susceptible to "ring bias," and with good reason. The appearance of Saturn and its rings through a backyard telescope is so breathtaking that it has to be shared with others. The "planet with rings" is one of the most popular targets for public star parties. Even professional astronomers tend to devote a great deal of attention to Saturn's rings. It's understandable when you consider that many an astronomer began a lifelong fascination of astronomy after viewing Saturn's rings through a telescope during his or her childhood.

Figure 6.1 Saturn. AP Photo/NASA.

There are other distractions that tend to turn our attention away from Saturn, the planet. When astronomers aren't dissecting the rings, they might be concentrating on those amazing moons whose stunning portraits are sent earthward by robotic probes. For now, however, let's imagine that Saturn has been stripped of its rings and moons and focus on the planet. We'll look at the rings later in this chapter and investigate the moons in Chapter 7.

EARLY OBSERVATIONS

When we view Saturn with the unaided eye, we see it the way early humans did. Saturn is a bright, but rather unspectacular object. It appears as a pale-yellow "star," shining anywhere between magnitudes −0.2 and +1.2. Saturn is brighter than all but a handful of nighttime stars, but it pales in comparison to the other naked eye planets. Compared to Venus (typically magnitude −4.5) and Jupiter (−2.5), Saturn appears almost feeble. Mars and Mercury also outshine Saturn on occasion.

Still, it wouldn't have taken much effort for our prehistoric ancestors to identify Saturn as the slowest moving of the five wandering "stars." As was the case with Jupiter, we'll never know the identity of the individual who discovered Saturn. By the time the earliest civilizations arose, Saturn was a recognized and watched heavenly body.

What distinguishes this wandering star from the others is its slow, deliberate movement relative to the background stars. To the Assyrians, it was Lubadsagush (oldest of the old sheep), perhaps because it appeared to amble through the zodiac like a grazing animal. The Greeks called it Kronus, the mythological father of Zeus—a name later changed to Saturnus by the Romans. Kronus was the god of agriculture (possibly a connection to the Assyrians' grazing sheep) and the keeper of human time. The latter is appropriate, because the 29.5 years this wanderer needs to complete a full cycle across the heavens was comparable to a typical human life expectancy in those days. Nowadays, we see Kronus annually without even realizing it. He's none other than Father Time, that white-haired, toga-wearing, old man who leaves the scene, scythe in hand, at New Year's!

The early skygazers correctly reasoned that if Saturn takes longer to traverse the zodiac than the other wanderers, it must be farther away from Earth. Even before the first telescopes were directed toward Saturn, astronomers had created models of the solar system that placed Saturn as the most distant of the planets.

SATURN: A MODERN VIEW

From the moment astronomers began pointing telescopes toward Saturn, the rings commanded center stage. Galileo led the way, but his crudely built telescope failed to reveal the rings in all their glory. He saw what appeared to be a pair of moons rigidly positioned on opposite sides of Saturn. Their disappearance within a few years' time left the Italian astronomer baffled.

The mid-1600s saw vast improvements in telescope design and optical quality. In 1655, the Dutch astronomer Christiaan Huygens discovered Saturn's largest moon, Titan. That same year, he correctly identified the true nature of Saturn's rings, but waited several years to formally announce the discovery. During the next three decades, Huygen's Italian counterpart, Giovanni Cassini, added four more moons and discovered the rift in Saturn's main rings that bears his name.

Larger and more sophisticated telescopes also meant astronomers could look past the rings and moons and concentrate on Saturn itself. Around the time that Huygens was discovering the true nature of Saturn's rings, the Italian astronomer Francisco Grimaldi reported that the planet's disk was slightly flattened from pole to pole. By the next century, more attention was being paid to Saturn's atmosphere, defined by pale-yellow bands running parallel to the equator. In 1794, a bright-white patch (a storm) appeared on Saturn's disk. The English astronomer William Herschel used it as a marker to make an early determination of Saturn's rotation period that was close to today's figure.

Once mathematicians worked out the proportions of the orbits of the planets and it was determined that Saturn circles the Sun at an average distance of 1,429,400,000 kilometers (857,640,000 miles, or 9.5 AU), Saturn's equatorial diameter could be determined. With an equatorial diameter discovered to be over nine times greater than Earth's, Saturn proved to be the solar system's second largest planet, after Jupiter. The orbital motions of Saturn's moons enabled astronomers to calculate Saturn's mass, and found it to be equal to 95 Earths. Considering Saturn's immense bulk (its volume is equal to 764 Earths), Saturn is actually something of a lightweight, with an overall density measured out at just 0.7 times that of water. Dropped into a vast ocean, the Ringed Planet would literally float!

The spectroscope and photographic plate, which appeared in the second half of the nineteenth century, revealed more of Saturn's secrets. Spectroscopic analysis of Saturn's atmosphere by the German astronomer Carl Vogel showed the presence of the hydrogen compounds ammonia and methane. With the photographic plate, more moons and a few faint rings came to light.

Our knowledge about Saturn took a Space Age quantum leap in 1979, when the first flyby by a manmade craft (*Pioneer 11*) returned close-up photos and data about Saturn's magnetic field. The *Voyager* missions of 1980 and 1981 analyzed Saturn's atmosphere, further mapped its magnetosphere, and provided stunning close-up views of Saturn, its intricate rings, and moons.

Two technological advances near the close of the twentieth century allowed space scientists to make observations of Saturn previously available only by robotic flybys like *Pioneer* and *Voyager*. The first was the Hubble Space Telescope, placed in orbit above Earth's often turbulent atmosphere.

Astronomers used the Hubble to observe Saturn's weather during times when no robotic craft were available to do the job. Hubble not only imaged Saturn in visible wavelengths, but also in the heat-generated infrared. Back on the ground, computer controlled adaptive optics were enabling earth-bound telescopes to automatically compensate for distortions caused by the atmosphere. Images of Saturn comparable to those produced by the Hubble were now possible from ground-based observatories. The CCD camera, able to collect faint light far more efficiently than the photographic plate, allowed astronomers to quickly image Saturn's atmosphere or search for faint moons.

By far the greatest tool for uncovering Saturn's secrets is one being used as this book goes to press. The *Cassini Orbiter*, a combined NASA and European Space Agency enterprise, arrived at Saturn in 2004 and conducted an extensive four-year survey. In 2008, due to its tremendous successes, the *Cassini* mission was extended for two more years.

How Can a Planet Float?

If the idea of a planet capable of floating in water seems hard to imagine, consider a typical iceberg. Even a small iceberg has the mass of a skyscraper. Out of water, it would crush anything beneath it. But the frozen water that comprises an iceberg is a rather lightweight material. If you placed a cubic centimeter of ice (a volume about half the size of a playing dice) on a scale, its mass would be just 0.8 grams. We would say its density is 0.8 grams per cubic centimeter (g/cm^3). An equal volume of room-temperature water would have a mass of exactly one gram and a density of $1.0 g/cm^3$. Saturn's overall density is just $0.7 g/cm^3$, slightly less than ice. That's because much of Saturn is made up of hydrogen and helium, the two lightest elements in the universe. Think of Saturn as the biggest iceberg in the solar system!

A SCALED-DOWN JUPITER

Saturn may be Jupiter's mythological father but, if we were to describe the relationship of the two planets from the viewpoint of their physical characteristics, it would be Jupiter's little brother. Jupiter's kid brother is still a giant. Its diameter is nine times greater than Earth's, and its volume could encompass over 760 Earth-sized planets. The Ringed Planet definitely deserves to be called a "gas giant."

Stripped of its rings, Saturn would appear Jupiter-like, with a slightly flattened disk, horizontally crossed by light and dark bands. The bands aren't as distinct and colorful as Jupiter's, and those near Saturn's equator are slightly wider. Saturn is nearly twice as far from the Sun as Jupiter, therefore its atmosphere is colder. Consequently, an upper layer of ammonia ice fog enshrouds the planet, muting the features beneath.

Saturn's atmosphere mirrors Jupiter's in chemical composition, being comprised mostly of molecular hydrogen (96.3 percent) and helium

(3.4 percent) by volume, with traces of methane, ammonia, and water. The latter substances produce the various shades of yellow seen in Saturn's banded atmosphere. A key difference between the two gas giants is that the percentage of helium in Saturn's outer atmosphere is lower than in Jupiter's.

Like Jupiter, Saturn radiates into space about twice as much energy as it receives from the Sun. While slow gravitational collapse explains this phenomenon for Jupiter, Saturn lacks the mass to produce this much internal energy by gravitational collapse alone. Noting that Saturn's outer atmosphere possesses a lower percentage of helium than Jupiter, astronomers speculate that the helium precipitates deep within Saturn's atmosphere. Being heavier than the surrounding hydrogen, the helium droplets "rain" downward into Saturn's interior. Friction from the falling helium could produce some of the heat energy detected.

SATURN'S ATMOSPHERE AND WEATHER

Mathematical and computer-generated models, based on the analyses of elements and temperatures found near Saturn's cloud tops, have led astronomers to speculate that Saturn's atmosphere is very Jupiter-like. One difference is the veil of ammonia haze that blankets Saturn—a result of the planet's colder temperatures, which hover around −140°C (−220°F) near the cloud tops. Beneath the haze, it's pretty much a repetition of Jupiter's atmosphere. Bright clouds hover above consecutive layers of ammonia ice, ammonia hydrosulphide crystals, water ice, and a water/ammonia fog.

Saturn's atmospheric content and banded appearance have already been mentioned. The cloud bands are similar to Jupiter's and are divided into zones (light) and belts (dark). Heat released from Saturn's interior combines with Saturn's rapid rotation to help generate the zones and belts.

The predominant wind belts run in the same easterly direction as Saturn's spin. In the early 1980s, the *Voyager* craft measured wind speeds near the equator in excess of 1,600 kilometers (1,000 miles) per hour. The *Cassini* mission has found the winds to be greatly reduced—only about 60 percent of what the *Voyagers* measured.

A windy atmosphere is also a stormy one, and Saturn has its fair share of storms. Most noticeable are the Great White Spots, which appear either in Saturn's northern hemisphere or close to the equator about once every Saturnian year. White spots were seen in 1933, 1960, and 1990, and were readily visible in backyard telescopes.

Storms have also been tracked in Saturn's southern hemisphere. In fact, an active zone between 30° and 35° south latitude has been nicknamed Storm Alley. As it approached Saturn in 2002, the *Cassini Orbiter* detected a pair of hurricane-like vortexes in Storm Alley. Each was about 1,000 kilometers (624 miles) across—about three times the diameter of Hurricane Katrina! The *Cassini Orbiter* detected storms in 2004 and 2006, each lasting about a month. In 2008, it tracked a monster storm spewing lightning bolts a thousand times stronger than those experienced in earthly thunderstorms. Saturn's storms appear to be created by enormous upwellings of warmer gas from Saturn's interior.

Totally unexpected atmospheric features have been found at each of Saturn's poles. An unusual hexagon-shaped cloud pattern, first found by the *Voyagers* and later confirmed by the *Cassini* mission, surrounds Saturn's north pole. The hexagon is 25,000 kilometers (15,000 miles) in diameter—wide enough to hold two Earths. It rotates in cadence with the planet's interior. In the autumn of 2008, the *Cassini Orbiter* was busy imaging Saturn's north pole in infrared light (this part of Saturn was immersed in winter darkness at the time). A huge cyclonic storm with winds twice as fast as the strongest earthly hurricane was found in the middle of the hexagon. Despite this activity, the hexagon remained unperturbed. The cause of this strange hexagon is presently unknown.

Another hurricane-like vortex, minus the hexagon, exists near Saturn's south pole; 8,000 kilometers (5,000 miles) across, it packs winds comparable to those found in the northern vortex. Clouds around the edge of this southern vortex are 30–75 kilometers (20–45 miles) higher than those near the center.

Robotic flybys like the *Pioneers* and *Voyagers* have sent back dramatic close-up images of Saturn, but there are drawbacks. The glimpses they provide of these faraway worlds are momentary at best. If we want to study slowly evolving climatic conditions, we need long-term surveillance. Weather patterns on the outer planets occur in seasonal cycles that occupy time spans measured in decades. Analyzing the dynamics of an outer planet's weather and climate with a flyby mission is like studying the ebb and flow of a basketball game by studying a few photos. An orbiter like *Cassini* will produce images covering several years, but the net product is still limited. It's like analyzing a basketball game using a five-minute video clip!

Prior to the earliest space probes to the planets, astronomers had to rely on ground-based telescopic observations. Distortions caused by air currents in our atmosphere permitted only the largest storms in Saturn's atmosphere to be followed. With the Hubble Space Telescope in concert with sensitive imaging equipment and computer-controlled adaptive optics, we are now able to monitor day-to-day weather and seasonal climate changes not only for Jupiter and Saturn, but for Uranus and Neptune as well.

SATURN'S INTERIOR

Like Jupiter, Saturn has no solid surface. Its interior is similar to Jupiter's. As we descend deeper into Saturn's hydrogen-rich atmosphere, increasing pressure liquefies the hydrogen gas, forming a slushy soup of liquid molecular hydrogen. Further down, where the pressure is one thousand times what we experience on the Earth's surface, the liquid molecular hydrogen then becomes liquid metallic hydrogen. In this state, hydrogen is a conductor of electricity, and it is the electric currents here that generate Saturn's magnetic field. At the very center of the planet, a hot, slushy layer of "ices" comprised of water, methane, and ammonia surrounds a core made up of silicate rock. Estimates place Saturn's core at 10 to 20 times Earth's mass, with a radius of 6,000 kilometers (3,700 miles). The temperature here may reach 12,000°C.

SATURN'S MAGNETOSPHERE

Saturn has a magnetic field whose source probably differs from Jupiter's. Jupiter's magnetic field seems to be generated by electrical currents formed in the spinning liquid metallic hydrogen zone deep in its interior. While Saturn's magnetic field may be created in a similar fashion, some astronomers believe that the helium rain condensing inside Saturn and sinking to the core may contribute by stirring up material in the liquid metallic hydrogen zone. Whatever its origin, Saturn's magnetic field is much weaker than Jupiter's. Like a bar magnet (not to mention Earth and Jupiter), Saturn's magnetic field is dipolar. What makes it unique in the solar system is its close alignment with the planet's axis of rotation. While Earth's magnetic field is skewed 11 degrees from its axis, Saturn's is less than a degree off.

Particles captured by Saturn's magnetic field help to form its magnetosphere. Like the Earth's and Jupiter's, Saturn's magnetosphere is shaped by the solar wind. On the sunward side of Saturn, the magnetosphere extends between 20 and 25 Saturn radii outward (depending on the strength of the solar wind). Away from the Sun, the magnetosphere extends at least 60 Saturn radii, well beyond the orbit of Titan.

A planet with an atmosphere and a magnetosphere is subject to auroral displays, and Saturn is no exception. Both the Hubble Space Telescope and the *Cassini Orbiter* have detected auroras in Saturn's polar regions. Unlike auroras here on Earth, whose durations are measured in hours, those on Saturn can last for several days.

....

Carolyn Porco (1955–)

When Carolyn Porco was 13 years old, a friend showed her Saturn through a rooftop telescope. It was a signature event for a young girl who would ultimately become the leader of the Imaging Team on the *Cassini* mission to Saturn. She studied astronomy at the State University of New York in Stony Brook, and then completed her doctorate at the California Institute of Technology.

While at Cal Tech, Dr. Porco got involved with NASA's Jet Propulsion Laboratory, helping to work on the data being returned from the *Voyagers'* Saturn encounters. She stayed on with the *Voyager* team through the Uranus and Neptune flybys, later applying for the *Cassini* mission position that she currently holds. Her communication skills have made Carolyn Porco a standby on numerous television news shows and documentaries.

....

A PLANET WITH JUG HANDLES

In 1610, several months after making his landmark observations of Jupiter's moons, Galileo turned his homemade refracting telescope in the direction of Saturn. We can only guess what thoughts ran through his mind when he peered into the eyepiece. Would Saturn be attended by orbiting moons like Jupiter? Would it be covered with craters like the Moon? Saturn turned out to be as surprising as Jupiter, but in a different way. In his notes, Galileo wrote that he could see "two small stars which touch it, one to the east and one to the west." He was aware that Saturn's puzzling appearance "results either from the imperfection of the telescope or the eye of the observer." The strange planet obviously required further observation.

Unlike Jupiter's moons, which wandered from one side of the planet to the other, Saturn's companions remained fixed when viewed on subsequent evenings. The Italian astronomer received an unexpected surprise when he revisited Saturn two years later. The moons were gone! So stunned was Galileo by this revelation, that he seriously questioned the reliability of his earlier sightings. "Has Saturn swallowed his children?" he wrote. Even though he predicted that the companions might return to visibility, he was still confused when, in 1613, the stars reappeared. This time, they were elliptical in shape, leading Galileo to describe them as "ears" or "jug handles," immovable and possibly attached to Saturn.

Galileo was correct in assuming that the imperfect optics of his telescope might be to blame for the odd appearance of Saturn and its mysterious companions. In 1655, using a telescope optically superior to Galileo's, the Dutch astronomer Christiaan Huygens discovered the true nature of Saturn's jug handles. Huygens noted, "It [Saturn] is surrounded by a thin, flat, ring, nowhere touching, inclined to the ecliptic." In his book *Systema Saturnium*, published four years later, he stated that the absence of Saturn's rings, as noted by Galileo in 1612, would occur whenever the Earth passes through the plane of the rings, presenting them edgewise to our line of sight. Being thin and flat, they would disappear in all but the most powerful telescopes. That the rings should be all but invisible even under telescopic scrutiny was proof that they must be incredibly thin!

In 1676, Giovanni Domenicus Cassini reported that the ring surrounding Saturn consists of two concentric parts separated by a gap (later called the Cassini Division). The outer of the two was denoted as the A Ring, and the inner as the B Ring. These are the bright rings that greet our eyes when we view Saturn with a telescope.

There were more rings surrounding Saturn, but they were too faint to be seen by the casual observer. In 1850, the astronomer William Bond and his son George were studying Saturn from the Harvard College Observatory. They noticed a dark band etched against Saturn's disk that appeared adjacent to the inner part of the B Ring. Could a faint, dusky ring reside inside

···

Giovanni Cassini (1625–1712)

At an early age, the Italian-born Giovanni Cassini became interested in astrology. His early astrological work led to a dual career in astrology and astronomy. From 1648–1669, Cassini worked at the Panzano Observatory, also teaching astronomy at the University of Bologna. Using unusually long refracting telescopes at Panzano, Cassini determined the rotation period of Mars and observed the gap in Saturn's rings that bears his name. He also discovered four new moons of Saturn.

In 1671, Cassini moved to France, accepting an invitation by King Louis XIV, the Sun King, to become Director of the Paris Observatory. He quickly adopted French citizenship, changing his name to Jean-Dominique. Perhaps his most significant achievement while there was a determination of Earth's distance from the Sun, using parallax measurements of Mars. His calculation was just 7 percent shy of the actual figure.

···

···

The Strange Telescopes of Huygens and Cassini

There was a reason Galileo had trouble getting sharp images from his telescopes. The convex front (objective) lenses in his refracting telescopes tended to bend light like a prism, disturbing the images. Telescope makers in subsequent decades tried to remedy the problem by using objective lenses that weren't as curved. This meant making longer telescopes. The tubes of some of these telescopes, which were used by Christiaan Huygens and Giovanni Cassini, were so long and cumbersome that they had to be mounted on tall towers. Fortunately, the introduction of the reflecting telescope in the late 1600s returned telescopes to a more manageable size and ended the era of the super-long refractors.

···

the B Ring? It could, and their discovery became known as the C Ring, or, because of its gossamer appearance, the "crepe ring."

The photographic plate allowed astronomers to detect even fainter rings. Photographs taken at the Allegheny Observatory in Pittsburgh, Pennsylvania, in 1966 and studied in detail by Walter Feibelman the following year revealed an extremely faint ring lying well beyond the A Ring. It would eventually be called the E Ring. Two years later, Pierre Guerin announced evidence pointing to a possible D Ring filling the void between Saturn and the C Ring.

Unmanned interplanetary space probes launched during the latter decades of the twentieth century produced an unparalleled look at Saturn's rings. *Pioneer 11*, launched in 1973, arrived at Saturn a half dozen years later. It confirmed the existence of Feibelman's E Ring and located the razor-thin F Ring between the E Ring and the bright A Ring. The *Voyager* probes launched in 1977 cruised past Saturn in 1980 (*Voyager 1*) and 1981 (*Voyager 2*). Not only did they uncover yet another ring (the G Ring, positioned between the E and F Rings), but they confirmed observations of unusual dark spokes in the B Ring reported by a handful of earthbound astronomers, most recently the amateur astronomer Stephen J. O'Meara. O'Meara made telescopic observations of the spokes several years before the *Voyagers* arrived at Saturn, but his reports had been generally discounted. The *Voyagers* also provided an amazing wealth of information about the complex structure of the rings in general. More ring gaps were discovered, and the A and B Rings were found to be comprised of thousands of concentric rings. Images of so-called empty regions like the Cassini and Encke Divisions betrayed the presence of material organized in the form of tiny ringlets.

The *Cassini Orbiter*, launched in 1997, arrived at Saturn in 2004. A back-lit photo of Saturn revealed two new rings related to some of Saturn's moons. The Janus/Epimetheus Ring was discovered between the F and G Rings, while the Pallene Ring was found to coincide with the orbit of the newly discovered moon, Pallene. Small ring arcs were reported around Saturn's tiny moons Atlas, Anthe, and Methone.

THE NATURE OF SATURN'S RINGS

What is the nature of Saturn's rings? It's a question astronomers have pondered from the moment in 1610 when Galileo observed "stars" on either side of Saturn to the present as we receive the latest images of the rings by the *Cassini Orbiter*. At first, astronomers thought they might be solid. In 1787, Pierre Simon LaPlace postulated that Saturn's rings were composed of many concentric solid ringlets. James Clerk Maxwell countered in 1859 that it would be impossible for the rings to be solid. His mathematical calculations showed that a solid ring would quickly break apart. Maxwell suggested that they might be made up of a multitude of small particles, each with its own orbit around Saturn. His idea was proved correct in 1895, when the astronomers James Keeler and William Campbell analyzed the rings with the spectroscope and found them to be rotating at different speeds. Just as planets close to the Sun orbit faster than those further out, so do ring particles nearest Saturn circle the planet more swiftly than those inhabiting the outer ring.

Radar and spectroscopic studies conducted in the twentieth century showed that Saturn's rings are made up of fragments ranging in size from microscopic dust grains to house-sized chunks. The fragments were found to be comprised of water ice or ice-covered rocky chunks with small quantities of space debris mixed in.

It's logical to wonder why the rings are so flat, and not torroidal (doughnut-shaped). The answer lies in the planet's equatorial bulge, which concentrates Saturn's gravity pull into a zone in space positioned directly above the equator. If any ring particle were to stray above or below the ring plane, it would quickly be pulled back into the zone. Each ring particle revolves around Saturn in an independent orbit, but we can imagine countless collisions, where particles bump together and fall out of orbit, only to be brought back in line by Saturn's gravity or the gravitational tug of a nearby moon.

Saturn's rings are amazingly thin—hardly more than a few dozen meters or yards in thickness. We see this in two different ways. Already mentioned is the disappearance of Saturn's rings when orbital conditions place them edge-on to our line of sight. Moreover, on those extremely rare occasions when Saturn occults (passes in front of) a bright star, the star continues to shine dimly through the rings. Imagine a scale model of Saturn where the rings have been reduced in size so they are now one kilometer (0.6 miles) across. Their thickness would be less than that of a sheet of cardboard!

Saturn's rings are incredibly beautiful, complex, and mysterious. They even have their own atmosphere. The *Cassini* mission detected trace amounts of oxygen surrounding the rings. The likely cause is the breakup of water ice contained in the ring particles.

THE ORIGIN OF THE RINGS

How did Saturn's rings come to be? In 1849, the French mathematician Edouarde Roche proposed that a fluid satellite had strayed too close to Saturn and was torn apart by that planet's gravity. He figured out a critical distance from a planet, known as the Roche Limit, at which a body of sufficient size and composition would be destroyed by the planet's gravity. An offshoot of Roche's theory held that the rings are part of the original solar nebula locked into a region of Saturn inside the Roche Limit and, therefore, unable to accrete into a moon. Yet another theory was that a Saturnian moon had been fragmented by a collision with a comet, the remains falling into orbit around Saturn.

The age of the rings is another puzzling question. Have they been around since Saturn's formation, or are they a relatively recent phenomenon, whose beauty our descendants in the distant future might not enjoy? Some astronomers argue that the brilliance of Saturn's rings, when compared to the drab rings of the remaining Jovian planets, points to a cosmically recent formation, perhaps within the past hundred million years. If the rings had formed at the same time as Saturn, 4.6 billion years ago, they would have long ago darkened as they became "polluted" with space dust.

Recent data from the *Cassini Orbiter* reveal that Saturn's main rings may be more massive than previously thought. That extra material would dilute space dust, retaining the rings' brightness. The grinding of ring particles exposes fresh material that also keeps the rings "fresh." New matter is added to the rings by Saturn's inner moons, whether from particles blasted off their surfaces by micrometeorites or (in the case of Enceladus) by water released from geysers. Our future descendants may not see Saturn the way we do, but there will be rings to delight their visual senses.

....

Now You See Them; Now You Don't

What would Saturn be like without its magnificent rings? Twice during Saturn's 29.5-year orbit of the Sun we get to find out. That's because Saturn's axis of rotation is tilted 27° to the plane of its orbit, causing the rings to occasionally appear edge-on to our line of sight. Such an event is known as a **ring plane crossing**.

Let's look at a complete cycle of Saturn's ring aspects, beginning with the ring plane crossing of 2009. Saturn's rings are edge-on to our line of sight and, because they are so incredibly thin, are virtually invisible in all but large, high-power telescopes. Astute observers notice a difference in Saturn's naked eye appearance. Without the extra light the highly reflective rings supply, Saturn appears dimmer than in years when the rings are "open." After 2009, the rings appear to open up as Saturn's northern hemisphere comes into view. They will be at their widest in 2017, allowing telescope users a nice view of detail in the cloud belts in Saturn's northern hemisphere. Saturn will also appear at its naked eye brightest.

Figure 6.2 Aspects of Saturn's rings. AP Photo/NASA/The Hubble Heritage Team.

As Saturn continues its stately trip around the Sun, the rings will appear to narrow until a second ring plane crossing in 2025, when they will again appear edge-on. For the next 14 years, we'll get a repeat performance of the dance of Saturn's rings, this time with the southern hemisphere tilted in our direction.

THE "A-B-C'S" OF SATURN'S RINGS

Saturn's rings are identified by letters from the alphabet. Because they were named in the general order of discovery, rather than by their distance from Saturn, we don't get a tidy A-B-C sequence when listing them in order from the surface of Saturn outward. Adding to the confusion is the fact that two new rings and three ring arcs discovered by the *Cassini* mission haven't been formally given letter designations. They remain named after the moons that spawned them. In order from the surface of Saturn outward, the rings are: D, C, B, A, Atlas ring arc, F, Janus/Epimetheus, G, Pallene, and E. Rings A, B, and C were the first three discovered, and form Saturn's main rings. The remainder are extremely faint, and are collectively referred to as the "dusky" rings. In addition, a trio of ring arcs has been found around Saturn's moons Atlas, Methone, and Anthe. Following is a summary of the characteristics of each.

D Ring

Saturn's innermost ring was suspected by the French astronomer Pierre Guerin in 1968 and confirmed by *Voyager 1* in 1980. The D Ring begins about 66,900 kilometers (41,570 miles) from Saturn's center, just 6,600 kilometers (4,100 miles) above its cloud tops, and extends outward for 7,500 kilometers (4,660 miles). It's extremely faint, and likely made of microscopic dust particles. Comprised of several tiny ringlets, the D Ring seems to be dynamic. When photos of the D Ring taken by *Voyager 1* were compared to photos taken by *Cassini* 25 years later, one of the ringlets in the D Ring was found to have moved 200 kilometers (125 miles) closer to Saturn.

An interesting aspect of the D Ring is its wavelike appearance, which was first noted by the Hubble Space Telescope in 1995 and more closely studied by the *Cassini* mission in 2006. Unlike most of Saturn's rings, which are thin and flat, the D Ring is punctuated by wavelike ripples a kilometer (one-half mile) in height. Individual ripples were separated by about 60 kilometers (37 miles) when imaged by the Hubble, but had closed to 30-kilometer (18.6-mile) separations when *Cassini* imaged them. One theory is that the waves were created when a comet swept through the D Ring sometime during the 1980s. Whatever the cause, the D Ring waves provide continuing proof that Saturn's rings are dynamic and ever-changing—even in the span of a human lifetime.

C Ring

One of Saturn's main rings, the C Ring was detected visually in 1850 by W. C. Bond and his son George at the Harvard College Observatory. Also known as the Crepe Ring because of its sheer appearance, the C Ring begins 74,650 kilometers (46,390 miles) from Saturn's center and continues outward about 17,500 kilometers (10,875 miles) half again as wide as the Earth. It is likely comprised of dust to pebble-sized particles.

Near the outer edge of the C Ring is a 270-kilometer (168-mile)-wide open space known as the Maxwell Gap. Discovered by the *Voyager* flybys in the early 1980s, it's named in honor of James Clerk Maxwell.

B Ring

The largest and most massive of Saturn's rings, the B ring (along with the A Ring) is what backyard astronomers see when they look at Saturn with modest-sized telescopes. Beginning 92,000 kilometers (57,170 miles) from Saturn's center, the B Ring is as wide as two planet Earths. It ends rather abruptly at the Cassini Gap. Comprised of particles of dirty water ice ranging in size from tiny grains to boulder-sized chunks, this rather dense, opaque ring is hardly more than several-dozen meters thick.

Mysterious Spokes

When *Voyager 1* and *2* reached Saturn in the early 1980s, they took detailed photos of the rings. Among the many surprises was the discovery of spokes in the B ring that appeared to rotate in unison with Saturn. They appear dark when backlit by the Sun and bright in forward-scattered light. The *Voyager* pictures confirmed observations made by amateur astronomers back on Earth.

The spokes appear to be a seasonal phenomenon, not visible during Saturn's summer or winter, but present near the equinoxes, when the rings are edge-on to the Sun. When the *Cassini* craft arrived at Saturn in 2004, the spokes were nowhere to be seen. A year later, they appeared as predicted. Because the spokes travel around the B Ring at the same rate as Saturn's axis spin, it is now believed that they are a phenomenon related to Saturn's magnetic field. As Saturn rotates, the magnetic field drifts through the rings, elevating microscopic particles and creating the "spokes."

....

Stephen James O'Meara (1956–)

"Eagle-eyed" is a term that best describes Stephen James O'Meara. In a day when photographic equipment is used to study the night sky, Stephen O'Meara is a throwback to a time when astronomers sat at the eyepiece and sketched what they saw. Regarded as one of the premier visual astronomers today, O'Meara boasts a resume to back up the claim. In 1980, he determined a rotation period for the planet Uranus by observing two bright clouds in its atmosphere. His reported value of 16.4 hours was in disagreement with the finding of professional astronomers who, using sophisticated equipment, were claiming a rotation period closer to 24 hours. Then *Voyager 2* arrived at Uranus and found a rotation rate just minutes different from O'Meara's. On January 24, 1985, he became the first person in the world to visually detect Halley's Comet prior to its 1986 rendezvous with the Sun, and was the last to see its departure.

But O'Meara's greatest visual triumph was probably an observation of the spokes in Saturn's B Ring. Studying the planet with a telescope in the mid-1970s, he noticed the spokes and reported that they moved at the same rate as Saturn's rotation. There was skepticism from the astronomical community. The B Ring rotates much slower than Saturn. How could the B Ring spokes move in unison with the planet? O'Meara was vindicated when the *Voyager 1* arrived at Saturn in 1980. Time-lapse movies made from the *Voyager* images proved that the spokes exist and do indeed circle Saturn in unison with the planet's rotation.

....

Cassini Division

Between Saturn's B and A Rings is a 4,700-kilometer (2,920-mile) gap named after its discoverer Giovanni Cassini, who reported it in 1676. The Cassini Division exists because particles orbiting in this zone are in a 2:1 resonance with Saturn's moon Mimas. Each time a particle in the zone orbits Saturn twice, it lines up with this moon. As a result, Mimas's gravity tugs the particle out of the zone. Telescopically, the Cassini Division appears dark and empty. Photographs from *Voyagers 1* and *2* showed that the Cassini Division is actually populated by a few tiny ringlets.

A Ring

The second of Saturn's bright rings, the A Ring begins at the outer edge of the Cassini Division 122,170 kilometers (75,915 miles) from the center of Saturn. More transparent and a little more than half as wide as the B Ring, it is also comprised of dirty water ice in particles ranging in size from tiny grains to chunks the size of soccer fields. In fact, the *Cassini Orbiter* has detected dozens of propeller-shaped wakes in the material of the A Ring that may be caused by tiny moonlets. Astronomers theorize that these "propeller" moonlets, which may number in the thousands, are constantly being built up and then torn apart by collisions with other particles in the ring.

There are two gaps in the A Ring. The Encke Gap, observed by Johann Encke in 1837, is a 325-kilometer (200-mile)-wide zone cleared out by Saturn's tiny moon Pan. The Keeler Gap, discovered by the *Voyagers* and named in honor of astronomer James Keeler, lies near the outer edge of the A Ring. It is a 35-mile (22-mile)-wide zone created by the moon Daphnis.

Atlas Ring Arc

In 2004, the *Cassini Orbiter* detected a faint ring between the A and F Rings. Designated R/2004 S1, it shares the orbit of Saturn's tiny moon Atlas. R/2004 S1 is either comprised of material removed from Atlas by micrometeorites or ring particles drifting inward and being captured in Atlas's orbit.

F Ring

The F Ring, discovered in 1979 by the *Pioneer 11* flyby, is arguably Saturn's most intricate ring. Known as Saturn's "braided" ring, it's only 30 to 500 kilometers (18.5 to 310 miles) wide and lies 140,180 kilometers (87,100 miles) from Saturn's center. Comprised of pebble-sized particles, the ring is corralled by two shepherd moons, Promethius on the inside and Pandora on the outside.

The *Cassini Orbiter* has provided close-up views of the F Ring, showing what appears to be a core ring with a second ring spiraling around it. *Cassini* photographs also show that Prometheus periodically gouges out channels in the F Ring whenever its orbit brings it nearby. This is arguably Saturn's most dynamic ring, with changes being observed on a daily basis.

Janus/Epimetheus Ring

Formed by minute particles ejected by micrometeorite impacts on the surfaces of Saturn's co-orbital moons Janus and Epimetheus, this extremely faint ring was discovered by the *Cassini Orbiter* in 2006. As we face Saturn with the Sun at our backs the way we view this planet from our vantage point on Earth, the Janus/Epimetheus Ring would be virtually invisible. From a position on Saturn's far side, with Saturn blocking out the Sun, the *Cassini Orbiter* was able to capture a backlit view of Saturn's rings. Sunlight passing through the Janus/Epimetheus Ring was scattered by its dusty material, betraying the ring's presence. The inner edge of the Janus/Epimetheus Ring lies 149,000 kilometers (92,590 miles) from Saturn, and is about as wide as the continental United States.

G Ring

Discovered by *Voyager 1* in 1980, the G Ring is another faint United States-wide ring. The inner edge of the G Ring is 170,000 kilometers (105,640 miles) from Saturn. It's unique among Saturn's rings in that its component particles, some of which are boulder-sized, are not scattered uniformly. They appear to be concentrated in a 250-kilometer (155-mile)-wide arc, which appears to be controlled by Saturn's moon Mimas. The *Cassini Orbiter* has found several other arcs, leading astronomers to speculate that the G Ring may be the remains of a destroyed moon. This ring may be slowly eroding away, bombarded by micrometeorites and dispersed by Saturn's magnetosphere.

Methone and Anthe Ring Arcs

The Methone ring arc, centered on Saturn's moon Methone, was detected by the *Cassini Orbiter*. The material in the arc is believed to represent dust ejected from Methone by micrometeoroid impacts.

Also discovered by the *Cassini Orbiter*, the Anthe ring arc co-orbits with and receives its material from the moon Anthe. Like the Methone ring arc, the Anthe ring arc appears to be controlled by a resonance with Saturn's moon Mimas.

Pallene Ring

Another discovery of the *Cassini Orbiter*, this 2,500-kilometer (1,550-mile)-wide ring is centered about 212,000 kilometers (131,740 miles) from Saturn. As was the case with the Janus/Epimetheus Ring, the Pallene Ring was observed in a backlit image taken while Cassini was on the side of Saturn opposite the Sun.

Orbiting within this ring is its namesake moon Pallene. Dust blown off this tiny moon by countless micrometeorite impacts is the probable source for its material.

E Ring

Saturn's remotest and by far widest ring, the E Ring was discovered photographically by Alan Feibelman at the Allegheny Observatory in 1967 and confirmed by *Pioneer 11* in 1979. So wide is this ring that it encompasses the orbits of five of Saturn's moons. It begins 181,000 kilometers (112,475 miles) from Saturn, near the moon Mimas, and continues past Enceladus, Tethys, and Dione all the way to Rhea—a span of 302,000 kilometers (187,660 miles).

Unlike most of Saturn's rings, which are exceedingly flat, the E Ring is doughnut-shaped. The source of its microscopic particles was discovered in a fascinating image returned to Earth by the *Cassini* mission. It clearly shows geysers of water erupting from the south polar region of Enceladus, spreading the material outward into the E Ring. We now know that the E Ring constantly receives new material from Enceladus, and that the gravity fields of Saturn's moons with an assist from Saturn's magnetic field distribute this material throughout the E Ring.

SEE FOR YOURSELF

Take one look at Saturn through a telescope, and you'll understand why a majority of astronomers place it at or near the top of their lists of must-see cosmic wonders. With the possible exception of the Moon, no other telescopic sight evokes more gasps of amazement than Saturn and its fabled rings. They are surprisingly easy to see, and may be glimpsed with a small backyard telescope magnifying just 20 times. From our earthly vantage point, we don't see Saturn from "above," so the rings don't appear as circles surrounding the entire planet. We view them at an angle, so they seem to cut in front of Saturn and disappear behind. Viewed with low magnifications, Saturn looks much like the flying saucers from a 1950s-era science fiction movie! The visibility of the rings depends on their angle relative to Earth. Twice during Saturn's 29.5-year orbit of the Sun, they appear edgewise to our line of sight and are virtually invisible—2009 and 2025 are years of such ring-plane crossings.

A low-power view with a small telescope will also capture Saturn's largest moon, Titan, which appears as a stellar speck. Nightly observations made over a few weeks' time will reveal Titan's 16-day orbit around Saturn.

With a medium-sized telescope and magnification of 100X, you can see Saturn's flattened disk, the Cassini Division in the rings, and a handful of Saturn's largest moons. Cloud belts can be faintly seen and, in rare instances, the occasional storm, visible as a bright white patch.

WEB SITES

http://saturn.jpl.nasa.gov/home/index.cfm.
    The official Web site for NASA's *Cassini-Huygens* mission.
http://sse.jpl.nasa.gov/planets/profile.cfm?Object=Saturn.
    NASA's informational site on Saturn.
http://www.astronomtcast.com/episode-59-saturn.
    The script to astronomers Fraser Cain and Pamela Gay's podcast on the planet
    Saturn. Includes links to the actual podcast and other Saturn-related Web sites.
http://www.nineplanets.org/saturn.html.
    Bill Arnett's Nine (8) Planets Web site is filled with facts about Saturn.

7

Saturn's Moons: A Death Star, a Deep Freeze Earth, and More

A VARIED FAMILY OF MOONS

Like Jupiter, Saturn is accompanied by a veritable swarm of moons—60 have been confirmed as of 2008. Many more may lie embedded in the rings, where astronomers have difficulty distinguishing between a large clump of ring particles and a tiny moonlet. Saturn's family of satellites also mimics Jupiter's in that many appear to be captured bodies, small in size and orbiting far from the main planet in highly inclined, retrograde orbits. Saturn's moons also tend to be comprised mostly of light "icy" material and are locked in synchronous orbits.

In size, the moons of Saturn are more varied than Jupiter's. Saturn is orbited by several spherical mid-sized moons that bridge the gap between one of planetary proportions (Titan) and a host of tiny irregularly shaped moons.

The first of Saturn's moons was discovered in 1655 by Christiaan Huygens. He named his find *Luna Saturni* (Saturn's Moon). Four more fell to the watchful eye of Giovanni Cassini between 1671 and 1684. He named them the *Sidem Lodoicea* (the Stars of Louis) in honor of King Louis XIV of France. In 1789, William Herschel christened a huge 48-inch reflecting telescope— the world's largest at the time—with the discovery of two more Saturnian moons. Like the one found by Huygens, they remained unnamed.

Herschel's son, John, took care of two problems—moons that were either anonymous or named after a king—by bestowing on Saturn's seven known moons the names of the Greek Titans, mythological brothers of Saturn. Huygen's moon became Titan, the four Cassini moons were renamed Tethys, Dione, Rhea, and Iapetus, and the elder Herschel's finds were designated Mimas and Enceladus.

An eighth satellite, Hyperion, was discovered in 1848 by the American father/son team of William and George Bond and, independently, by the English astronomer William Lassell. Hyperion would be the last of Saturn's moons to be discovered visually. Approximately 50 years passed before Saturn's ninth moon, Phoebe, was found on a photographic plate made by W. H. Pickering.

Throughout the first half of the twentieth century, Saturn's moon count remained at nine. Then, in December 1966, came the announcement of a possible tenth moon, Janus. Two astronomers independently claimed discovery: Audouin Dollfus at the Pic du Midi Observatory in France and Richard Walker at the Flagstaff Observatory in Arizona.

The *Voyager* space probes ushered in a new era of satellite discovery. Analysis of the images they made during their 1980 and 1981 flybys brought to light eight new moons, all associated with Saturn's rings. Two of these moons proved to be the co-orbitals Janus and Epimetheus, observed as a single moon by Dollfuss and Walker in 1966. More recently, the *Cassini Orbiter*, which arrived at Saturn in 2004, found three more and added a trio of unconfirmed moons in the F Ring.

As was the case with Jupiter, the discovery of the majority of the satellites attending Saturn came thanks to the technological advances of CCD photography and adaptive optics, which greatly enhance the performance of Earth-based telescopes. Using the wide-field Supreme Cam affixed to the giant 8.2-meter (26.7-foot) Subaru Telescope on Mauna Kea in Hawaii, a team of astronomers, led by Scott Sheppard and David Jewitt, unearthed dozens of Saturnian satellites between 2004 and 2007. Most are tiny moons with highly inclined, retrograde orbits.

The satellites of Saturn may be loosely lumped into three groups: moons and moonlets associated with Saturn's rings, eight large "middle" moons, and a host of moonlets orbiting far from Saturn.

"PROPELLER" MOONLETS

In 2006, the *Cassini* probe imaged four propeller-shaped wakes in Saturn's A Ring. They were attributed to disturbances in the ring material caused by tiny moonlets approximately the size of soccer fields. Since then, dozens of "propeller" moonlets have been uncovered, and they may number in the thousands.

If a body the size of a soccer field were discovered in orbit around Earth, it would immediately be hailed as our second Moon. For now, the propeller

Figure 7.1 Pan. NASA.

moonlets are simply considered to be large chunks of ring matter, rather than separate moons.

THE SHEPHERD MOONS

The closest recognized moons to Saturn are the shepherd moons, or ring shepherds—so-called because they orbit in and around Saturn's rings and gravitationally shepherd, or "herd," the particles. Some, like Pan (shown in Figure 7.1) and Daphnis (DAF-nis) plow through the rings, creating gaps. Atlas, Prometheus (pra-MEE-thee-us), and Pandora (pan-DOR-uh), on the other hand, orbit just outside Saturn's rings, defining their sharp edges.

All of the Shepherd moons are relatively small and irregularly shaped. Like the ring particles they control, they orbit Saturn in a **prograde, or direct, motion**.

JANUS AND EPIMETHEUS, CO-ORBITALS "TRADING SPACES"

Judged by orbital characteristics alone, Saturn's co-orbitals Janus (JAY-nus) and Epimetheus (ep-eh-MEE-thee-us) would be considered among the most amazing moons in the solar system. They literally trade places! The pair share an orbit just outside the F Ring, a mere 91,000 kilometers

Figure 7.2 Janus and Epimetheus. NASA.

(56,550 miles) above Saturn's cloud tops. Only 50 kilometers (30 miles) separate their orbits, with the closer of the two to Saturn orbiting slightly faster. Every four years, when the faster-moving moon catches up to its partner, they exchange momentum and switch places, the outer moon moving inward as the inner moon shifts to the outer orbit. After doing a cosmic equivalent of square dancing's "do-si-do," they go their separate ways until the next encounter/switch four years later.

In 1966, the French astronomer Audouin Dollfus announced the discovery of a tenth moon of Saturn (later named Janus) just three days before the American astronomer Richard Walker also claimed discovery. Twelve years later, the American astronomers Stephen Larson and John Fountain proposed that Janus might actually be a pair of Saturnian moons. Their findings were confirmed when *Voyager 1* captured both Janus and Epimetheus in 1980. Today, Dollfus is usually given credit for the discovery of Janus, while Walker, Larson, and Fountain are considered codiscoverers of Epimetheus.

Janus is irregular in shape, its longest dimension about 200 kilometers (120 miles). Its surface is ancient and heavily pockmarked with craters up to 30 kilometers (20 miles) across. Epimetheus is about three-fourths as large as Janus. Like its partner, Epimetheus has an ancient, heavily cratered surface. Because of their shared orbits, astronomers theorize that Janus and Epimetheus may have once been part of a single body that broke apart early in the formation of Saturn's moon system.

MIMAS, THE "DEATH STAR" MOON

Pronounced: "MY-mass"
Diameter: 392 km (244 mi)
Mean Distance from Saturn: 185,520 km (115,280 mi)
Orbital Period: 0.94 days
Period of Rotation: 0.94 days (synchronous)
Average Density: 1.2 g/cm^3
Surface Gravity, Compared to Earth: 0.7%

Show a *Voyager* photograph of Mimas to a *Star Wars* fan, and he or she will immediately respond, "It looks like the Death Star!" The moon's uncanny resemblance to the evil Empire's planet-destroying space vessel is due to the presence of a huge crater created by an impact that nearly shattered Mimas. Named Herschel after Sir William Herschel, who discovered this moon in 1789, the crater is 130 kilometers (80 miles) wide—one-third the diameter of Mimas. Herschel Crater is 10 kilometers (6 miles) deep and is punctuated by a central peak 6 kilometers (4 miles) high. If a crater of these

Figure 7.3 Mimas. AP Photo/NASA.

proportions existed on Earth, it would be as large as the continent of Australia, with a central peak 205 kilometers (128 miles) high.

The surface of Mimas is littered with craters, many with diameters exceeding 40 kilometers (25 miles). An exception is a region near Mimas's south pole, where few craters wider than 20 kilometers (12 miles) exist.

Like our own Moon and many of the planetary moons in the solar system, Mimas has a synchronous orbit, with one side perpetually locked toward Saturn. Nearest of Saturn's mid-sized moons, Mimas circles its home planet in a little less than one Earth day.

Synchronous orbits seem to be the rule for Saturn's main moons, as is an icy composition. With an overall density of 1.2 g/cm^3, Mimas is only slightly denser than water. It's understandable that Mimas would be frozen. Its surface temperature has been measured at $-200°$C ($-328°$ F).

Because of Mimas's proximity to Saturn's rings, its gravity controls their appearance. Mimas is the prime cause for the Cassini Gap that separates Saturn's A and B Rings. Particles orbiting in this zone revolve around Saturn in what astronomers refer to as a 2:1 resonance. When a ring particle circles Saturn twice, it lines up with Mimas and gets a gravitational tug that pulls it out of the Cassini Gap. Mimas is also responsible for the boundary between Saturn's B and C Rings (a 3:1 resonance), and it has a 3:2 resonance with the F Ring shepherd moon Pandora.

METHONE, ANTHE, AND PALLENE—PIECES OF MIMAS?

Three tiny moons, Methone (mi-THOH-nee), Anthe (AN-thee), and Pallene (Pal-LEE-nee), orbit between Mimas and Saturn's next moon outward, Enceladus. They were discovered between 2004 and 2007 by the *Cassini* imaging team, led by Carolyn Porco. Astronomers believe they may be fragments of Mimas or Enceladus. It's also possible that all five of these moons are remnants of an even larger moon.

ENCELADUS, A WATER FOUNTAIN?

Pronounced: "en-SELL-ah-dus"
Diameter: 494 km (306 mi)
Mean Distance from Saturn: 238,020 km (149,905 mi)
Orbital Period: 1.37 days
Period of Rotation: 1.37 days (synchronous)
Average Density: 1.6 g/cm^3
Surface Gravity, Compared to Earth: 1.2%

If you look at some of the basic data on Saturn's moon Enceladus, you might assume that it's a carbon copy of Mimas. It's slightly larger, has a similar synchronous orbit around Saturn, and has an ice-laden surface.

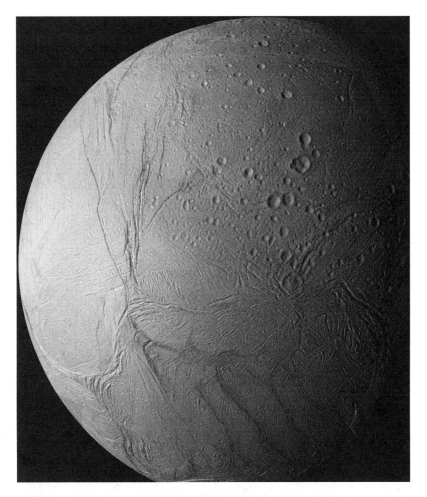

Figure 7.4 Enceladus. AP Photo/NASA Jet Propulsion Laboratory/Space Science Institute.

Enceladus and Mimas even share the same discoverer, William Herschel, and year of discovery, 1789.

The similarities end there. The surface of Enceladus is twice as reflective as Mimas's, with an albedo of 0.99, meaning it returns 99 percent of the sunlight that strikes it. For comparison, freshly fallen snow sports an albedo between 0.80 and 0.90. There is no gigantic "Death Star" crater on Enceladus. In fact, Enceladus is relatively crater free. Its smooth surface, interrupted by long cracks and ridges, indicates a relatively young, active surface.

This was proved when the Cassini orbiter photographed enormous geyser-like plumes emanating from Enceladus's south pole. The geysers were seen to spew material hundreds of kilometers into space, some of it being swept into Saturn's E Ring. Enceladus is now believed to be the prime source of material for that ring, providing evidence that the formation of at least some of Saturn's rings is an ongoing process.

The geysers originate from a series of large, dark fissures nicknamed "tiger stripes." While much of this moon's surface is locked in a deep freeze of −201°C (−330°F), the temperature in the vicinity of the tiger stripes is a "balmy" (−93°C, or −135°F). In March 2008, the *Cassini* probe skimmed the surface of Enceladus, coming as close as 48 kilometers (30 miles) and actually drifting through one of Enceladus's icy geysers. A "taste" of its material showed the presence of water vapor and **organic compounds**. In a NASA press release, one *Cassini* scientist compared Enceladus's brew to "carbonated water with an essence of natural gas."

Planetary scientists must now consider the possibility that reservoirs of liquid water lie just beneath the surface of Enceladus. Whether the heat source for Enceladus's water geysers is tidal flux from Saturn's gravity and resonance with the moons Tethys and Dione, or from an internal heat source, is still being debated. The fact that the source of the organic compounds detected by *Cassini* exists somewhere in the interior of Enceladus makes this moon a candidate, though remote, for some form of extraterrestrial life. Enceladus will continue to receive plenty of attention from the *Cassini* mission, and will be a prime focus for any future probes sent to Saturn.

TETHYS, A GIANT ICE BALL

Pronounced: "TEE-thiss"
Diameter: 1,066 km (662 mi)

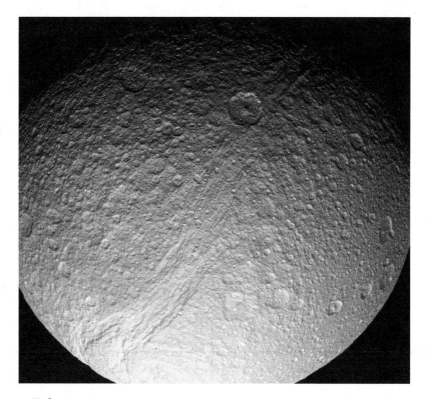

Figure 7.5 Tethys. NASA.

Mean Distance from Saturn: 294,660 km (183,100 mi)
Orbital Period: 1.89 days
Period of Rotation: 1.89 days (synchronous)
Average Density: 1.0 g/cm^3
Surface Gravity, Compared to Earth: 1.5%

Tethys was discovered by Giovanni Cassini in 1684. A lightweight among Saturn's major moons, Tethys has a mean density of just 1.0 g/cm^3. This would indicate that Tethys is largely composed of ices with some rocky material present in the core.

Like Mimas, Tethys is dominated by a huge crater, the result of a cosmic impact that nearly shattered it. The crater, Odysseus, spans 400 kilometers (250 miles), more than one-third the diameter of Tethys. In contrast to the well-defined features associated with Mimas's "Death Star" crater Herschel, the walls and central peak of Odysseus are somewhat flattened.

Another of Tethys's more interesting features is the Ithaca Chasma. This gigantic fracture zone, 100 kilometers (62 miles) wide and 3 to 5 kilometers (2 to 3 miles) deep, runs through the poles, roughly two-thirds of the way around the moon's circumference. Some astronomers conclude that Ithaca Chasma exists because the surface of Tethys froze first and then cracked when the interior later froze and expanded. Others, noting that Ithaca Chasma is located opposite Odysseus Crater, believe that the two were created by the same impact.

Tethys is co-orbited by a pair of moons, Telesto (tah-LESS-toh) and Calypso (ka-LIP-so), each about 35 kilometers (22 miles) across. Telesto and Calypso are positioned approximately 60 degrees to either side of Tethys in stable orbital locations known as **Lagrange points**. Tethys is in a 1:2 **orbital resonance** with another Saturnian moon, Mimas.

DIONE, A SPUN-AROUND MOON?

Pronounced: "dy-OH-nee"
Diameter: 1,123 km (685 mi)
Mean Distance from Saturn: 377,400 km (234,520 mi)
Orbital Period: 2.74 days
Period of Rotation: 2.74 days
Average Density: 1.5 g/cm^3
Surface Gravity, Compared to Earth: 2.3%

Dione and Tethys are twins in size, and both were discovered by Giovanni Cassini in 1684. As was the case with like-sized Mimas and Enceladus, the similarities between Dione and Tethys end quickly. Dione is half again as dense as Tethys, hinting at a rocky silicate core that accounts for one-third of its total mass, the remainder being ice.

Figure 7.6 Dione. AP Photo/NASA Jet Propulsion Laboratory/Space Science Institute.

Studies of Dione's surface have baffled astronomers. Locked in a synchronous orbit around Saturn, Dione's leading edge should be more heavily cratered as it plows headfirst into oncoming space debris. This is not the case. It's Dione's trailing edge that bears the scars of impacts with large objects. Many of these craters have diameters greater than 100 kilometers (62 miles). Taking into account Dione's relatively small size, astronomers believe any object massive enough to leave behind craters that large would literally spin Dione around. It's possible that billions of years ago, Dione was struck by such an object, turning its heavily cratered leading hemisphere to the rear. Considering the large number of outsized craters dotting Dione's surface, such spin-arounds could have repeated several times. Eventually, Dione would have settled into its current synchronous orbit. Noting that Dion's leading hemisphere is slightly darker than the one that trails behind (a result of an accumulation of micrometeorite dust), astronomers estimate that Dione has been "locked in" for several billion years.

Besides craters, Dione has some bright, wispy features on its surface. These may have resulted from eruptions of icy material in Dione's distant past, as no activity resembling Enceladus's geysers has been seen.

Dione mimics Tethys by sharing its orbit with a pair of smaller moons. Helene (HEL-uh-nee) and Polydeuces (POL-i-DEW-seez) are both situated in stable positions at Lagrange points 60 degrees to either side of Dione.

Helene, whose greatest diameter is about 35 kilometers (22 miles) is 10 times larger than tiny Polydeuces.

As far from Saturn as our Moon is from Earth, Dione nevertheless zips around Saturn 10 times faster than our Moon orbits Earth. Dione is in orbital resonance with Enceladus, having an orbital period exactly twice Enceladus's.

RHEA, A RINGED SATELLITE?

Pronounced: "REE-a"
Diameter: 1,528 km (949 mi)
Mean Distance from Saturn: 527,040 km (327,490 mi)
Orbital Period: 4.52 days
Period of Rotation: 4.52 days (synchronous)
Average Density: 1.2 g/cm^3
Surface Gravity, Compared to Earth: 2.7%

The second largest of Saturn's moons, Rhea was discovered by Cassini in 1672. At first glance, Rhea appears to be a larger edition of Dione, having a

Figure 7.7 Rhea. NASA.

synchronous orbit (in this case 4.5 days), similar composition, and comparable surface features. Temperatures on Rhea's surface mimic those found on both Tethys and Dione, ranging from −174°C (−281°F) in sunlight to −220°C (−364°F) in shaded areas.

Rhea has dissimilar leading and following edges like Dione, but doesn't appear to have been "spun around." Unlike Tethys and Dione, whose surface features seem "watered down" by resurfacing, Rhea exhibits an ancient landscape with sharply defined features. While rocky material seems to be concentrated near the cores of Tethys and Dione, it seems to be evenly distributed throughout Rhea's interior. It's as if Rhea cooled off too quickly after its formation to allow heavy matter to settle in its center.

Cassini images of Rhea show two major types of geologic terrain, based on crater density. Much of this moon is littered with craters, many of which are larger than 40 kilometers (25 miles) in diameter. Craters in parts of the polar and equatorial regions of Rhea aren't as large, indicating a past episode of resurfacing. Evidence of this is found in the light, wispy marks that cover parts of Rhea's surface. Similar in appearance to those found on Tethys and Dione, they are revealed by high-resolution photos to be fractures tens to hundreds of kilometers in length.

Late in 2005, planetary scientists studying data relayed back to Earth from the *Cassini* spacecraft made a startling discovery. A detecting device on board the *Cassini* craft that measures the electron flow from Saturn's magnetosphere was scanning the area around Rhea in search of a tenuous atmosphere. The number of electrons reaching the detector did indeed appear to drop, but not at a steady uninterrupted pace. Instead, there was a near-symmetrical pattern of three sharp drop-offs to either side of Rhea. Something surrounding Rhea was blocking the paths of the electrons. A preliminary explanation is that three faint rings or ring arcs containing a sparse population of pebble- to boulder-sized particles surround Rhea. The three potential rings lie between 1,600 and 2,100 kilometers (1,000 and 1,300 miles) from Rhea's center. Saturn has always been known as the Ringed Planet. Could one of its moons be the Ringed Satellite?

TITAN, AN "EARTH" IN DEEP FREEZE

Pronounced: "TIE-tan"
Diameter: 5,150 km (3,200 mi)
Mean Distance from Saturn: 1,221,830 km (759,210 mi)
Orbital Period: 15.95 days
Period of Rotation: 15.95 days (synchronous)*
Average Density: 1.9 g/cm³
Surface Gravity, Compared to Earth: 13.8%
* Varies slowly over time, due to a slow drifting of Titan's crust

An entire chapter of this book could be devoted to Saturn's aptly named moon Titan. This huge sphere of rock and ice dwarfs Saturn's remaining moons and contains 90 percent of all the mass surrounding Saturn. With an equatorial diameter of 5,150 kilometers (3,200 miles), Titan is larger than the planet Mercury. Of all the planetary satellites in the solar system, only Jupiter's Ganymede is greater in size. Even mighty Ganymede falls short of Titan in one important area—atmosphere. Titan is unique in that it is the only moon in the solar system known to support a substantial atmosphere.

Discovered in 1655 by the Dutch astronomer Christiaan Huygens, Titan is readily visible in small backyard telescopes as an 8th magnitude star that circles Saturn once every 16 days. From then until 1944, Titan was written off as an ordinary speck of light. That's when spectroscopic analysis of its light showed that the little speck is surrounded by a dense atmosphere. Among the gases detected in Titan's atmosphere was methane, which is also found in the atmospheres of all four Jovian planets.

The *Voyager* flybys in 1980 and 1981 showed just how extensive Titan's atmosphere is. In images sent back to Earth by the *Voyagers*, Titan was found to be completely veiled by an orange-colored haze 300 kilometers (200 miles) deep. Besides methane, the *Voyagers* detected a large amount of the heavy gas nitrogen. A nitrogen-rich atmosphere should exert reasonably high pressure on Titan's surface. Sure enough, the air pressure at Titan's

Figure 7.8 Titan. NASA.

Figure 7.8 Continued

surface was found to be 1.6 bars (one bar equals Earth's atmospheric pressure at sea level). An astronaut standing on Titan would feel a squeeze similar to what he or she would experience at the bottom of a swimming pool. However, standing unprotected on Titan's surface would not be an option. Temperatures there, as measured by the *Voyagers*, are a frigid $-178°C$ ($-289°F$). Methane exists as a liquid at this temperature. Could Titan's surface be covered in an ocean of liquid methane? Unable to see Titan's surface, astronomers could only speculate. The giant moon was in no mood to readily reveal its deepest secrets!

With no space probes scheduled to explore the outer solar system for at least two decades, astronomers had no choice but to resort to studying Titan with Earth-based instruments. A bad situation took a turn for the better in the

1990s with the development of computer-assisted adaptive optics. Better yet, the Hubble Space Telescope was launched into Earth orbit high above our turbulent atmosphere. With bigger and better telescopic eyes, astronomers got a tantalizing glimpse of a world that indeed seemed to have areas of icy high elevation interspersed with large, flat expanses that might be bodies filled with some sort of liquid substance. Still, Titan seemed to be a remote, alien world.

Prior to the *Cassini-Huygens* mission, any discussion of which body in the solar system is most earthlike (Earth excluded, of course) centered on the planet Mars. Here is a world whose day is just a half hour longer than ours. Its axis tilt is similar, creating the Martian seasons. Temperatures are much colder and the atmosphere is thinner, but Mars has water, even if it is in a frozen state beneath the surface. The Martian terrain, whose canyons, gullies, and sand dunes were obviously etched by the erosive forces of wind and (in the past) water, is earthlike.

Thanks to the *Cassini-Huygens* mission, Mars will likely have to relinquish its "Most Earthlike" title. Taking images in near-infrared light while mapping with radar, the *Cassini Orbiter* was able to penetrate Titan's orange smog and flesh out surface detail. Even better, the *Huygens* probe descended into Titan's atmosphere, taking pictures before and shortly after soft-landing on the surface. Together, the two painted a portrait of a planetary satellite that might better be described as Earth in a deep freeze.

In 1994, the Hubble Space Telescope took a near-infrared image of Titan that showed what appeared to be a large hilly landmass the size of Australia. Christened Xanadu, it seemed to be surrounded by a flat region, possibly the surface of a large body of water.

A decade later, radar on board the *Cassini Orbiter* was able to peel away Titan's persistent smog and dissect Xanadu's topography, spotlighting a landscape that is surprisingly earthlike. Large expanses of dunes, broken by hills and valleys, were revealed. *Cassini* also detected possible winding river channels that led into level regions interpreted to be large lakes or lake beds. On a world too cold for flowing water, the liquid in Titan's rivers and lakes had to be methane or ethane.

Overall, the *Cassini Orbiter* pieced together a picture of a moon with a dense nitrogen-methane atmosphere hiding a relatively smooth, young surface. Impact craters are few and far between, and there is evidence of volcanoes that might spew a water-ammonia mix or methane.

Because of the presence of methane in Titan's atmosphere, astronomers postulated that Titan might be covered with oceans of the substance. Their reasoning lay in the fact that methane is broken down by contact with the Sun's ultraviolet radiation. Without a new source, the methane currently in Titan's atmosphere would be depleted within 50 million years—a cosmic blink of the eye.

Instead of discovering a global methane-ethane ocean, *Cassini* found a number of large lakes in Titan's northern hemisphere and a smattering in the southern hemisphere. The equatorial region (investigated by the

Huygens probe) proved to be rather dry. At the time of the *Cassini-Huygens* mission, Titan's northern hemisphere was in the throes of winter. Astronomers surmise that the slightly cooler temperatures promoted methane-ethane rainfall, which accumulated in the lakes.

Mapping Through the Haze

The *Voyagers* were stumped by the smoggy shroud that enveloped Titan. The *Cassini Orbiter* wasn't. Tipped off by the *Voyagers* that Titan was veiled by a dense atmosphere, the designers of the *Cassini* mission made sure their orbiter was fixed with instruments that could penetrate the fog.

The first was a near-infrared imager, similar to the ones used with Earth-based telescopes and the Hubble. Near-infrared, which can pass through Titan's cloudy atmosphere the way visible light cannot, maps the various kinds of surfaces by analyzing the sunlight they reflect.

The *Cassini Orbiter* also used radar to map Titan's surface. Radar is similar to the sonar used by sailors to navigate through uncharted waters. A sound wave sent downward through the water strikes the bottom and echoes back. The deeper the water, the greater the amount of time for the sound wave to hit bottom and return. Radar is a similar technique, with radar waves replacing sound. A radar scan can also provide information about how smooth a surface is. Hilly terrain tends to look bright, while flat surfaces appear dark. Prior to *Cassini's* investigation of Titan, radar was used by other probes to map the surface of cloud-bedecked Venus.

Near-infrared and radar provide an indirect view, but the best way to get a clear picture of Titan's surface is to land there. That's what the *Huygens* probe did. On January 14, 2005, having separated from the *Cassini Orbiter* 20 days earlier, *Huygens* sailed into Titan's dense atmosphere. Slowed by friction, then a parachute, the probe analyzed Titan's atmosphere while imaging the surface below.

The photographs taken by *Huygens* as it descended showed that smooth areas previously thought to be methane-ethane lakes were actually vast dune fields. *Huygens* landed in a relatively flat area littered with rounded pebbles and covered with what behaved like moist sand or gravel laced with organic compounds. On the deep freeze that is Titan, water ice behaves like rock, forming the sand and pebbles that cover its surface. The liquid that eroded the pebbles and made the soil moist was methane.

"How's the weather?" That's not an appropriate question for an astronaut stationed on our airless Moon, but it certainly works for Titan. It's the only moon in the solar system with a substantial atmosphere and the potential for a future job opening for a meteorologist. The conditions aren't as violent as what we find in Saturn's stormy atmosphere. There are no gigantic lightning bolts tearing through the sky. Unlike the Ringed Planet's supersonic winds, Titan experiences a mild global breeze from the slightly cooler poles toward the equator. Titan's typical day-to-day weather can best be described as "cold and drizzly." Combined with the gloom of a sky that blocks most of what little sunlight reaches Titan's cloud tops, and we have some pretty dismal weather.

As Titan's northern hemisphere switches from winter into spring and summer (seasons last about seven years), astronomers will be watching closely to see if the currently observed weather patterns switch and the southern hemisphere becomes the "Land of Lakes."

In 2007, Titan produced yet another surprise for astronomers. Data collected over a period of several years by *Cassini's* radar spacecraft showed that landmarks on Titan's surface had shifted position by as much as 30 kilometers (20 miles). This is possible only if Titan's icy crust is afloat on a vast internal ocean, possibly consisting of a water-ammonia blend. The cause of the wandering crust is likely Titan's global winds, which create friction with the crust and slowly move it.

Preliminary models of Titan's interior, based upon this discovery, envision an icy crust 100 kilometers (62 miles) thick floating on a 400-kilometer (250-mile)-deep water-ammonia ocean. Beneath that, a 200-kilometer (125-mile)-thick sheath of ices under high pressure surrounds Titan's 3,750-kilometer (2,345-mile)-wide rocky core.

Fill'er up? In these early years of the twenty-first century, we face a global fuel crisis. Limited supplies of oil and natural gas, both hydrocarbons, are steadily dwindling. On Titan, methane and ethane (prime ingredients in natural gas) exist in abundance. According to a 2008 NASA news release, Titan has hundreds of times more hydrocarbons than all the known oil and natural gas reserves on Earth. Could Titan be the site of future refineries, or will we have developed alternative fuel and energy technology to the point where we can leave Titan in its pristine condition?

Could life evolve on Titan? Has it already taken a foothold? Scientists speculate that Titan, with its nitrogen-methane atmosphere and organic-covered surface, replicates a "prebiotic" Earth, that is, Earth before its first life forms evolved. It's much colder out there, but life might have begun in an area warmed by a nearby volcano. Perhaps a future expedition to this "Earth in Deep Freeze" will provide the answer.

Christiaan Huygens (1629–1695)

The Dutch mathematician, astronomer, and physicist Christiaan Huygens was born in The Hague on April 14, 1629. From 1645 to 1649, he studied law and mathematics at the University of Leiden and the College of Orange at Breda before turning to a career in science.

In 1654, Huygens devised an improved method of lens grinding. The following year, he used one of his telescopes to discover Saturn's largest moon, Titan, and make observations of Saturn's rings that led to a realistic description of their true nature.

From 1665 to 1681, Huygens worked at the Paris Observatory. It was during his tenure there that the Dutch scientist made one of his greatest contributions to science—publicizing the idea that light travels in waves. The *Huygens* probe (part of the *Cassini-Huygens* mission to Saturn) that soft-landed on Titan in 2005 was named in his honor.

HYPERION, A COSMIC SPONGE?

Pronounced: "hi-PEER-ee-en"
Diameter: irregular—410 X 260 X 220 km (255 X 163 X 137 mi)
Mean Distance from Saturn: 1,481,100 km (920,300 mi)
Orbital Period: 21.28 days
Period of Rotation: undefined (varies)
Average Density: 0.6 g/cm^3
Surface Gravity, Compared to Earth: 0.2%

It's a honeycomb! It's a sponge! No, it's Saturn's moon, Hyperion! Perhaps the oddest-looking of Saturn's moons, Hyperion may truly be more honeycomb or sponge than moon. To begin with, Hyperion's numerous craters don't have the familiar, classic bowl-shaped form. One of Hyperion's hemispheres is dominated by a single huge crater 100 kilometers (62 miles) across and 10 kilometers (6 miles) deep. In and around this crater are numerous deep craters that look more like bottomless pits or openings to caves. With an overall density only half that of water, Hyperion may well be a huge mass of porous ice, honeycombed by caverns.

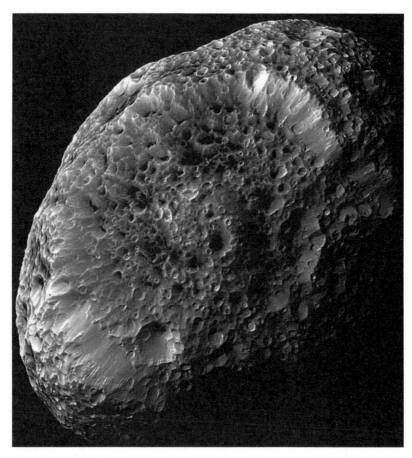

Figure 7.9 Hyperion. AP Photo/NASA.

Hyperion is the largest nonspherical planetary satellite in the solar system. It's possible that Hyperion was larger and more spherical in the distant past and was blasted into its present form by a cosmic collision. You might expect a moon comprised primarily of ices to be bright and highly reflective. On the contrary, Hyperion appears to be overlain with a thin carpet of reddish material, possibly cast down by Saturn's moon Phoebe.

Odd not only in shape and appearance, Hyperion also has a period of rotation that astronomers simply cannot define. While most of Saturn's large moons are locked into synchronous orbits, revolving and rotating at the same rate, Hyperion tumbles chaotically around its home planet. It's possible that Hyperion's odd shape causes it to twist each time it gets a gravitational tug from Saturn. Hyperion's 3:4 orbital resonance with Titan may be creating yet another gravity tug each time it passes near the giant moon.

Even the circumstances surrounding Hyperion's discovery are unusual. It was discovered independently in September 1848 by William. C. Bond and his son George P. Bond in the United States, discoverers of Saturn's C (Crepe) Ring, and by William Lassell in England.

IAPETUS, A TWO-SIDED MOON

Pronounced: "eye-AP-i-tus"
Diameter: 1,471 km (914 mi)
Mean Distance from Saturn: 3,561,300 km (2,213,000 mi)

Figure 7.10 Iapetus. AP Photo/NASA.

Orbital Period: 79.33 days
Period of Rotation: 79.33 days (synchronous)
Average Density: 1.2 g/cm^3
Surface Gravity, Compared to Earth: 2.2%

Hyperion doesn't have a monopoly on weird. Long before the first space probe arrived at Saturn to study this strange moon close-up, astronomers knew Iapetus was unusual. Giovanni Cassini, who discovered Iapetus in 1671, could only see the moon when it was on the west side of Saturn. He guessed that one side of Iapetus must be much darker than the other, and that the moon orbits with one side perpetually pointed toward Saturn. Later astronomers, using larger, more light-sensitive telescopes than his, were able to see Iapetus when it was to Saturn's east. However, it appeared 10 times fainter. Had Cassini painted an accurate portrait of this hide-and-seek moon?

He had! Over three centuries later, the *Voyager* flybys confirmed that Iapetus, indeed, has a synchronous orbit and that one hemisphere is noticeably darker than the other. An area on Iapetus's leading edge, dubbed the Cassini Regio, is as dark as charcoal, reflecting just 4 percent of the light striking it. By contrast, the trailing side of Iapetus is extremely reflective. As it orbits Saturn, Iapetus turns its dark hemisphere earthward when it's on the easterly side of Saturn, turning its bright hemisphere toward us when it drifts to Saturn's west. Iapetus is literally a two-faced moon.

The explanation for these extremes is still debatable. One theory holds that the leading edge of Iapetus is picking up dust and debris from space, possibly material from Saturn's dark moon Phoebe. The problem with this idea is that the dark material doesn't quite match Phoebe's color. Also, the floors of some of Iapetus's craters contain dark material, leading another camp of theorists to guess that it came from the moon's interior.

A popular hypothesis for the existence of Iapetus's starkly different sides is that the surface material on the dark side absorbs sunlight, slightly raising its temperature and causing water ice beneath to evaporate. The evaporated water drifts to the back side of Iapetus, where it freezes to the surface and brightens it.

Iapetus isn't just weird, it's downright nutty—literally! The *Voyagers* discovered a vast ridge stretching 1,300 kilometers (800 miles) along Iapetus's equator and rising 10 kilometers (6 miles) above the surrounding landscape. Called the Voyager Mountains, the ridge gives Iapetus a walnut-like appearance. Astronomers are still debating whether this mysterious feature is a folded mountain belt, a ridge of material that spewed out through a crack in Iapetus's surface, or even the remnants of a ring that collapsed.

PHOEBE, OUTERMOST OF SATURN'S MAJOR MOONS

Pronounced: "FEE-bee"
Diameter: 220 km (132 mi)

Figure 7.11 Phoebe. AP Photo/NASA Jet Propulsion Laboratory.

Mean Distance from Saturn: 12,952,000 km (8,049,668 mi)
Orbital Period: 548.0 days (retrograde)
Period of Rotation: 0.39 days
Average Density: 1.6 g/cm^3
Surface Gravity, Compared to Earth: 0.5%

In 1899, William H. Pickering found a new satellite of Saturn on a photo-graph—the first to be discovered photographically. Phoebe was found to be in a highly inclined retrograde orbit four times farther from Saturn's than any of the Ringed Planet's previously known satellites.

Not much was known about Phoebe until the *Voyager 2* encounter in 1981. Even then, the moon was 2.2 million kilometers (1.4 million miles) away as the probe cruised by. Its dark appearance in the *Voyager* images, combined with its highly inclined retrograde orbit led astronomers to surmise that Phoebe might be a captured asteroid.

The *Cassini* probe brought an end to the captured asteroid theory. Passing a thousand times closer to Phoebe than *Voyager 2*, it found Phoebe to be a primordial mix of ice, rock, and carbon-containing compounds. Astronomers now think Phoebe is made of the same compounds as objects found in the Kuiper belt. Because it may be a captured Kuiper belt object (KBO) and comprised of material left over from the formation of the solar system, Phoebe is of great interest to planetary scientists.

THE OUTER SATELLITES

Phoebe is by far the largest of Saturn's outer satellites. Between 2000 and 2007, Earth-based telescopes with adaptive optics and sensitive CCD cameras have brought to light dozens of moons in a region far removed from Saturn and its inner family of satellites. Most were discovered by a team from the University of Hawaii, headed by Scott Sheppard. The group used a wide-field CCD detector with the 8.2-meter (26.7-foot) Subaru Telescope perched atop Hawaii's Mauna Kea.

Like the outer satellites of Jupiter, Saturn's remote moons have highly inclined orbits and are most likely captured objects. A majority have retrograde orbits. None, except for Phoebe, has a diameter greater than 40 kilometers (25 miles). Saturn's outer moons can be categorized in three broad groups, based on orbital characteristics.

Inuit Group

The Inuit Group consists of five small moons, all discovered since 2000 and named after characters in Inuit mythology. They orbit a zone between 11,294,800 and 17,910,600 kilometers (7,018,600 and 11,129,600 miles) and have orbital inclinations between 46° and 50°. Unlike the majority of Saturn's outer moons, the members of the Inuit Group have direct (prograde) orbits with periods between 15 and 30 months. None has a diameter greater than 40 kilometers (25 miles).

Norse Group

A large collection of highly inclined retrograde satellites, the Norse Group includes Phoebe and a collection of 28 much smaller moons. Their orbits lie between 12,952,000 and 24,504,900 kilometers (8,049,700 and 15,227,300 miles) at inclinations between 137° and 177°. The nearest,

Phoebe, circles Saturn once every 18 months, while the most remote, the tiny 6-kilometer (3.7-mile)-wide Fornjot (FOR-nyot) takes nearly four years to complete an orbit.

Except for Phoebe, all of these satellites bear names taken from Norse mythology and were discovered from 2000 to the present. Only Phoebe has a diameter greater than 20 kilometers (12 miles).

Gallic Group

These four small moons, all discovered between 2000 and 2004 and named after figures in Gallic mythology, have orbits and compositions so similar that astronomers believe they are the remains of a larger object that broke apart. The Gallic moons have prograde orbits with periods ranging from 26 to 31 months. Found between 16,266,700 and 18,563,000 kilometers (10,107,700 and 11,535,000 miles) from Saturn, the Gallic moons have orbits with inclinations between 34° and 41°.

SATURN'S MOONS: THE MOST POPULAR FUTURE TOURIST SITES

In Chapter 5, we put a futuristic spotlight on a few vacation spots tourists will be flocking to when travel to Jupiter's moons becomes feasible. The moons of Saturn offer yet more adventures for twenty-second-century space travelers. Here are a few "can't miss" tourist sites.

1. Mimas. Jupiter viewed from Amalthea is impressive. Saturn seen from Mimas is absolutely breathtaking! Our lounge on Mimas Base faces Saturn, offering a spellbinding panoramic view of the Ringed Planet. In one Earth day, you'll experience a complete orbit around Saturn, seeing the planet and its magnificent rings in all their glory. Twice we pass through the ring plane, affording an extraordinary edge-on view of these razor-thin marvels of nature. While on Saturn's night side, we'll glimpse flashes of light from powerful thunderstorms and use special UV detecting equipment to glimpse the planet's beautiful auroras. Honeymooners, why go to Niagara Falls, when you can come to Mimas and enjoy the romance of Saturn!

2. Enceladus—the "Tiger Stripe" Geysers. So you think nothing can match the spectacle of an eruption of Old Faithful Geyser in Yellowstone National Park? You obviously haven't experienced the "Tiger Stripe" Geysers on Saturn's moon, Mimas. Old Faithful spits water a few hundred feet in the air; the Tiger Stripe Geysers blast it into outer space! Arrange a special Rainbow Tour, scheduled for those times when the Sun is at the right angle to the

Tiger Stripe Geysers. You'll be treated to one of the most spectacular rainbows in the solar system!

3. Titan. Come fly with us on Titan Tours! You won't be traveling in one of those old-fashioned prop-driven airplanes or fuel-guzzling jets your great-grandparents used on Earth a century ago. Instead, you'll realize humankind's greatest dream—you'll fly with your own wings! Thanks to Titan's dense atmosphere and low gravity, you can actually strap on a pair of our specially designed wings and fly like a bird! Soar peacefully over Titan's alien, but serenely beautiful terrain. See firsthand the methane lakes discovered by primitive robotic probes at the beginning of the twenty-first century. Join us on Titan Tours for an honest-to-goodness bird's-eye view of Saturn's biggest moon.

4. Hyperion. Calling all spelunkers! If you enjoy cave exploring, come to the Hyperion Caves. Billions of years ago a cosmic collision nearly pulverized Hyperion, leaving behind a vast network of caves that honeycomb its interior. If you're an older cave explorer who easily fatigues on a walk through Carlsbad Caverns, you'll delight in a Hyperion Caves expedition. Under Hyperion's low-gravity environment, you'll be light as a feather! Let our experienced guides take you on an inside tour of one of Saturn's most amazing moons.

WEB SITES

http://solarsystem.nasa.gov/planets/profile.cfm?Display=Moons&Object=Saturn.
　　Look here for NASA's look at Saturn's moons.
http://www.astronomtcast.com/episode-61-saturns-moons.
　　Script from "Astronomycast" podcast on Saturn's moons. Includes links to actual podcast and Web sites that provide updated information.
http://www.dtm.ciw.edu/sheppard/satellites/satsatdata.html.
　　Scott Sheppard's planetary satellite Web site. Updated regularly.
http://www.nineplanets.org/saturn.html.
　　Includes data on some of Saturn's major moons.
http://www.ted.com/index.php/talks/carolyn_porco_flies_us_to_saturn.html.
　　An inspirational video clip of a talk about the *Cassini* mission studies of Saturn's moons Titan and Enceladus. Presented by *Cassini* Imaging Team leader Dr. Carolyn Porco.

8

Uranus, a Tipped-Over World

January 24, 1986: Since early November of the previous year, *Voyager 2* has been keeping a robotic eye on the planet Uranus. *Voyager 2* is now racing 81,500 kilometers (50,600 miles) above the planet's bluish cloud tops. To compensate for the low-light conditions and *Voyager*'s great speed, scientists have programmed its cameras to lock onto each target as the craft speeds by. *Voyager 2* sends images of Uranus, its rings, and moons back to Earth, presenting astronomers with astonishing, never-before-seen close-up views of the mysterious planet.

A landmark moment of discovery would quickly be forgotten. Just four days after the *Voyager 2* encounter with Uranus, the Space Shuttle *Challenger* would blow up after takeoff, taking the lives of all seven astronauts on board. A nation in mourning will have little interest in a planet billions of kilometers from Earth. Michelle is saddened by the news, but she continues planning for her seventh-grade science fair project. She will prepare a poster board display that describes the planet Uranus. Her report will include *Voyager 2* findings and photographs relating to Uranus. It will contain more scientific facts than were ever known by the planet's discoverer, William Herschel.

Uranus Data

Period of Revolution	84.0 years
Period of Rotation	$17^h\,14^m$
Axis Tilt	98°
Equatorial Diameter	51,100 km (31,750 mi)
Mass (Earth = 1)	14.5
Surface Gravity* (Earth = 1)	0.9
Density (water = 1.0 g/cm³)	1.27
Number of Moons	27
Mean Distance from Sun	19.2 AU (2,871,990,000 km [1,784,950,000 mi])

* Since none of the Jovian planets has a solid surface, gravity is calculated at the visible cloud tops.

A SERENDIPITOUS DISCOVERY

In the unimaginably bleak and cold depths of the solar system, over a billion kilometers (six hundred million miles) beyond Saturn, a giant moved slowly and deliberately on its 84-year voyage around the Sun. So far from

Figure 8.1 Uranus. AP Photo/NASA.

Earth that it was barely visible to the unaided eye on the clearest and darkest of nights, this phantom drifted unnoticed through the background stars.

It had been seen, but not recognized. In 1690, the English astronomer Sir John Flamsteed mistook it for a faint star in the constellation Taurus. He was compiling a star catalog and mistakenly catalogued the faux star as 34 Tauri. In the mid-1700s, another Englishman, James Bradley, noted it several times but failed to recognize its significance. On the other side of the English Channel, the French astronomer Charles Le Monnier sighted it a dozen times. Like Flamsteed and Bradley, he thought it was a star. It was moving too slowly to be captured by a cursory glance. That situation was about to change.

The year was 1781. The American colonies were waging a war of independence against England. Los Angeles was founded by a group of settlers from what is now Mexico. Wolfgang Amadeus Mozart introduced his opera *Idomeneo* and Violin Sonata, K. 379. The solar system was known to harbor six planets—Mercury, Venus, Earth, Mars, Jupiter, and Saturn. It had been that way since the dawn of recorded history.

On the evening of March 13, a little-known German-English musician for whom astronomy was a hobby turned his telescope heavenward. Unlike his predecessors, William Herschel was a methodical observer who plied the starry sky with surgeon-like precision. Herschel didn't make random forays into the night sky, as had been the practice of all astronomers since the time of Galileo. He surveyed the constellations methodically, one star at a time. It was his intention to find and catalogue all of the double and multiple stars, star clusters, and nebulae that his telescope could reveal.

And what telescopes they were! Herschel was a master craftsman, whose homebuilt reflecting telescopes were the largest and finest on Earth. One of them, a telescope with a wooden tube 2 meters (7 feet) in length, housing a metal alloy mirror 15 centimeters (6 inches) in diameter, was Herschel's instrument of choice on this particular night.

While making his way through the stars of the constellation Gemini, Herschel chanced upon an unusual "star." Unlike the typical star that is so far away it appears as a pinpoint of light even when scrutinized with the highest magnifications, this "star" had a tiny, but distinct, bluish disk. Subsequent observations betrayed its motion against the background stars. Herschel reported his find, assuming he had uncovered a comet. Alerted by Herschel's announcement of a moving object in Gemini, astronomers watched closely. To calculate its orbit, they meticulously recorded its changing positions. What they found was an orbit too circular and remote for a comet. The strange object that William Herschel had discovered was a new planet orbiting the Sun far beyond Saturn. On that late winter night in 1781, an amateur astronomer had doubled the size of the known solar system.

To honor his king, Herschel christened his new planet "*Georgium Sidus*" (George's Star)—a name somewhat unpopular outside of England. The name "Herschel" was suggested, then "Uranus," in keeping with the mythological nature of planetary identities.

Serendipity is loosely defined as finding one thing during a quest for something else. A prospector who stumbles upon a deposit of diamonds while seeking gold has made a serendipitous discovery. Herschel's discovery of Uranus while searching for double stars, star clusters, and nebulae ranks as one of the most serendipitous finds in the annals of astronomy.

···

William Herschel (1738–1822)

German-born Friedrich Wilhelm Herschel (born in Hanover) first worked as a musician. An oboist in his father's Hanoverian Guards band, he left Germany while in his late teens and immigrated to England. Ultimately settling in the city of Bath, he soon established himself as a reputable musician, music teacher, composer, and conductor. He also Anglicized his first name to William.

While in his mid-30s, Herschel developed an interest in astronomy that quickly turned into a lifelong passion. Unsatisfied with the telescopes made available to him, he learned the art of telescope-making and was soon studying the heavens with reflecting telescopes that were the world's largest and finest.

On March 13, 1781, while undertaking an exploratory survey of the heavens, William Herschel came upon the planet Uranus. The discovery would gain him instant fame and a lifelong pension from King George III. From now on, Herschel was free to work full-time as an astronomer.

Besides discovering Uranus and two of its moons, William Herschel found hundreds of star clusters and nebulae, provided the first evidence that many double stars are true gravitationally bound systems, made the earliest observation-based map of our Milky Way Galaxy, and determined the direction of the Sun's motion in the galaxy. He also made landmark observations of the Sun, Moon, and planets. Astronomy historians today generally regard William Herschel as the greatest observational astronomer of all time.

William Herschel wasn't the only member of his family to earn fame as an astronomer. His sister Caroline (1750–1848) left Germany to join him in England, where she worked long hours as his assistant. On her own, she discovered several star clusters and nebulae, as well as six comets. William Herschel's son John (1792–1871) continued his father's sky survey into the southern skies by establishing an observatory in South Africa. A grandson, Alexander Stewart Herschel (1836–1907), studied meteors and the relationship between comets and meteor showers.

An interesting astronomical anecdote links William Herschel to his planet. When Herschel died in 1822, he was 84 years old, the same number of years it takes the planet he discovered to orbit the Sun.

···

In 1787, Herschel discovered two moons of Uranus. The odd tilt of their orbits indicated that Uranus must likewise be tilted. Its axis is now known to be tipped 98 degrees to the plane of its orbit around the Sun. Put another way, there are times during its 84-year cycle of the Sun that Uranus literally rolls across the sky!

Once the relative spacing of the planets was known and astronomers had accurately determined Earth's distance from the Sun, it was a simple mathematical matter to establish the distance to Uranus. Knowing this, astronomers could calculate the planet's diameter, which would lead to a determination of its volume. Though smaller than either Jupiter or Saturn, Uranus was still shown to be four times larger in diameter and 63 times greater in volume than Earth.

Meanwhile, the motion of its moons would betray Uranus's mass. With its mass and volume in hand, astronomers could now calculate the planet's overall density. Piece by piece, a picture began to emerge of a huge, but not very dense, world. Like Jupiter and Saturn, Uranus was an imposing body comprised of lightweight materials.

Other than these basic physical characteristics, astronomers knew little about Uranus. A planet so remote is reluctant to yield its secrets. Its tiny disk, equivalent in apparent size to a golf ball placed two kilometers (1.3 miles) away, yielded precious little information. Three more moons were discovered—two by William Lassell in 1851, and another on a photograph taken by Gerard Kuiper in 1948. Astronomers were able to analyze its atmosphere with the spectroscope. The predominant gases detected were methane and molecular hydrogen.

In 1977, almost two centuries after William Herschel's serendipitous discovery of Uranus, two teams of astronomers, one on the ground and one in an airborne observatory, made another accidental find. A predicted **occultation** of a bright star by Uranus offered astronomers a unique opportunity to study the composition and characteristics of Uranus's atmosphere as the planet passed in front of the star. To their surprise, the star began blinking on and off *before* Uranus occulted it. The pattern repeated after the occultation. The conclusion was that Uranus must be surrounded by a thin and very faint system of rings. Saturn was no longer the solar system's only ringed planet.

On New Year's Day, 1986, astronomers knew that Uranus was a big, gaseous planet that orbited the Sun once every 84 years, was tilted on its side, and was the focal point of a system of faint rings and five orbiting moons. In a matter of weeks, *Voyager 2* would rewrite the books on Uranus. During its Uranus flyby, *Voyager 2* obtained the first-ever close-up photos of Uranus's disk, revealing a rather bland surface. The 5 known moons were imaged, and 11 new ones were found. A photograph of Uranus's ring system added 2 more.

Voyager 2 also studied the composition, temperature, and wind speeds in the atmosphere of Uranus and mapped the planet's magnetic field. The spin of the magnetosphere, tied closely to the planet's rotating interior, gave astronomers a clue to the exact rotation period of Uranus, previously known with little certainty. The data and images assembled by *Voyager* would take space scientists years to analyze.

Once the diameters and masses of Uranus and its near-twin Neptune were accurately known, astronomers realized that Uranus might be the solar system's third-largest planet in diameter after Jupiter and Saturn, but it ranks fourth in mass, with Neptune being third. Having more mass and, therefore, a stronger gravitational field, Neptune is more compressed than Uranus.

As *Voyager 2* departed for Neptune and with no future plans by NASA to send further probes to the outer reaches of the solar system, astronomers had little choice but to resume studying Uranus from ground-based observatories. Fortunately, technology came to the rescue. The 1990s saw the launch of the Hubble Space Telescope, the development of adaptive optics, and the emergence of the Charge-Coupled Device (CCD). With bigger and keener "eyes," astronomers have discovered two more rings and nine previously unknown moons, and continue to study the changing weather patterns in Uranus's atmosphere.

BIRTH OF AN "ICE GIANT"

Until recently, all four of the outer planets were referred to as gas giants. However, there is a distinct difference between the compositions of Jupiter and Saturn (true gas giants) and Uranus and Neptune. The latter two seem to fill a niche somewhere between gas giants and the small, rocky terrestrial planets.

Jupiter and Saturn are comprised almost entirely of hydrogen, first in their gaseous outer atmospheres, and then in a liquid state in their interiors. By contrast, the outer atmospheres of Uranus and Neptune have higher percentages of methane (an "ice") in their hydrogen-dominated outer atmospheres. Below this is a hot, soupy interior made up of "ices" like water, methane, and ammonia, possibly surrounding a rocky core. While Jupiter and Saturn are almost totally made up of hydrogen and helium, Uranus seems to contain about 15 percent hydrogen and a small amount of helium, with the rest being mostly a rock/ice mixture. In a sense, Uranus might be similar to the cores of the two giant planets, minus their extensive outer shells of liquid hydrogen.

In Chapter 3, we noted that Uranus might have migrated outward from the site of its formation in the solar nebula. Theorists, backed by computer models, explain that Uranus could not have formed in its current location. At the time of the solar system's birth, they note, this region of the solar nebula didn't contain enough small bodies to accrete into a Uranus-sized planet. Computer models do show that Uranus could have formed in the same dense region of the solar nebula as Jupiter and Saturn, and then was gravitationally flung outward to its present orbit.

The disk instability model of the solar system's birth (also mentioned in Chapter 3) takes a different approach. Uranus might have formed quickly in its current location as gases in the outer part of the solar nebula gathered into spiral-shaped masses. One of these masses collapsed to form Uranus. As is always the case in science, new data will help astronomers determine which theory is true.

Another question relating to Uranus's early history centers on the planet's lopsided axis tilt. Most astronomers believe that during the chaotic birth of the solar system Uranus was impacted by a smaller protoplanet and wound up tipped on its side.

URANUS: THE SOLAR SYSTEM'S MOST BORING PLANET?

Voyager 1 and *2* sent us remarkable and totally unexpected snapshots of Jupiter and Saturn. Who could forget those images of Jupiter's turbulent Red Spot, Io's erupting volcanoes, and Saturn's majestic rings? As *Voyager 2* approached Uranus in late 1985, astronomers waited with high anticipation for the first up-close images of this remote world. They would be keenly disappointed. Instead of a stormy, banded atmosphere like the ones that envelope Jupiter and Saturn, Uranus presented a pale-blue, featureless face that seemed bland and downright boring. Computer enhancements revealed about 10 clouds, a far cry from the hundreds that littered the atmospheres of Jupiter and Saturn.

Why was Uranus so much more sedate that its big brothers? Data forwarded by *Voyager 2* on the amount of heat energy released by Uranus offered a clue. Whereas Jupiter and Saturn give off about two times as much heat energy as they receive from the Sun, Uranus shows little difference between the inflow of solar energy and the outflow of heat energy. Since this internal heat seems to be the driving mechanism in the weather patterns on Jupiter and Saturn, Uranus's relatively serene atmosphere seemed to make sense.

We now know that Uranus was being coy. At the time of *Voyager 2*'s Uranus encounter, the planet was in the middle of winter in the northern hemisphere, with the southern half of the planet (in summer) pointing almost directly at the Sun. In the two decades since then, as Uranus headed toward a December 2007 equinox, the Sun began to shine on the northern hemisphere. As it did, the atmosphere of Uranus began to stir. Images taken with the Hubble Space Telescope and the huge ground-based Keck Telescope in Hawaii showed the emergence of huge clouds, some nearly Earth-sized. Uranus was coming to life!

It's tempting to attribute human characteristics to the planets. Jupiter might be characterized as the king—regal, brash, and boastful. Saturn could be the handsome, vain prince. Uranus reminds us of a person who, at the beginning of a social engagement, sits quietly. Just when we typecast this individual as boring, he/she begins to liven up, eventually carrying on animated conversations with other guests. Like that initially quiet guest, Uranus was never boring—just a little bit shy. It's a wallflower among planets!

HOW'S THE (URANIAN) WEATHER?

Weather on Uranus is certainly different from what we experience here on Earth. First of all, you wouldn't want to step out for a breath of fresh air unless you don't mind inhaling an atmosphere comprised of 83 percent molecular hydrogen, 15 percent helium, 2 percent methane, and a pinch of other gases like ammonia. The methane content in its atmosphere is one reason why Uranus differs from Jupiter and Saturn, whose atmospheres are almost entirely hydrogen and helium.

Because it's 19 times farther from the Sun than Earth, Uranus is much colder. On Earth, the lowest surface temperature ever recorded was $-89\degree C$ ($-129°F$) at the Russian Vostok Station in Antarctica. Your skin, directly exposed to this cold, would begin to freeze in seconds. Breathing would be painful and result in damage to the lungs. Now imagine what it would be like to be exposed to the atmosphere of Uranus, where temperatures have been measured at $-216°C$ ($-357°F$). Unlike the Earth, where the coldest temperatures occur at the poles, Uranus's temperature is pretty uniform from pole to pole.

Its remote distance from the Sun isn't the only reason Uranus is so cold. We already noted that Uranus radiates about the same amount of heat energy into space as it receives from the Sun. The other Jovian planets all release about twice the heat energy they receive, indicating an internal heat source that slightly warms their atmospheres. What's wrong with Uranus? Perhaps it has no internal heat source, or escaping internal heat is somehow trapped by Uranus's atmosphere. Astronomers can only guess.

Uranus may have a docile appearance, but its winds still exceed the Earth's strongest hurricanes. Wind speeds and direction vary with latitude. Like Jupiter and Saturn, the atmosphere of Uranus is arranged in belts running parallel to the equator. Unlike the two gas giants, winds at the equator run from east to west in a direction opposite the planet's rotation. In that way, they are like Earth's equator-hugging trade winds, only colder and much faster. Equatorial winds as high as 400 kilometers (250 miles) per hour have been measured. At mid-latitudes, the winds run west-to-east at speeds of 720 kilometers (450 miles) per hour—more than twice the speed of Earth's fastest hurricanes.

Because Uranus's atmosphere, like those of Jupiter and Saturn, is much deeper than Earth's, we would find several layers of clouds. The uppermost is a layer of hydrocarbon fog comprised mainly of acetylene and ethane. Further down, clouds made up of different substances exist in discreet layers. At the top is a methane cloud deck. It is the methane that gives Uranus its distinct bluish-green hue. Methane absorbs red light, leaving mostly blue and green light to exit Uranus. Astronomers believe that beneath the methane cloud deck are clouds of ammonia or hydrogen sulfide, ammonia hydrosulfide, and water vapor.

During the early years of the twenty-first century, as spring began to arrive in Uranus's northern hemisphere, astronomers kept a close watch on the planet. Using large telescopes and adaptive optics, they saw increased activity in the atmosphere of what had been considered a sedate planet. Massive cloud systems, some almost Earth-sized, have begun to crop up. Because Uranus doesn't seem to have an internal heat source to drive its weather the way Jupiter and Saturn do, and sunlight at this distance is too weak to create strong weather patterns, the cause of this activity is as yet unknown.

"THE FOUR SEASONS" AS VIVALDI NEVER IMAGINED

You might be familiar with Antonio Vivaldi's beautiful musical composition *The Four Seasons*. Most of us have experienced the annual climate changes associated with the seasons his music portrays so elegantly. On Earth, as on the other planets, the seasons are created by a planet's axis tilt. For Earth, that tilt amounts to 23.5 degrees. Around the time of summer solstice in the northern hemisphere, the axis tilts this part of the Earth sunward. People who live in northerly latitudes enjoy a greater amount of daylight hours and bask in the Sun's direct rays, while populations south of the equator are experiencing winter's short days and slanted sunlight. Half a year later as winter grips the northern hemisphere, the opposite is true. In-between are the equinoxes, when the Sun shines directly above the equator and both hemispheres receive an equal dose of sunlight.

Now imagine what the seasons must be like on a planet whose axis tilt is 98 degrees, and whose orbit around the Sun takes 84 years. The equinoxes on Uranus would be like Earth's, with both hemispheres receiving an equal amount of sunlight. But the solstices would be a different matter! Imagine being in a spaceship holding a permanent position above the cloud tops on Uranus's north pole. At the time of spring solstice in the north, the Sun would make its first appearance above the horizon. As spring progressed, the Sun would appear higher and higher in the Uranian sky. After a spring that lasts 21 years, summer solstice would place the Sun as high in the sky as possible. For the next 21 years, the Sun would seem to migrate back toward the horizon until the autumn equinox, when it again lies on the horizon. It might have been nice to see sunlight for a continuous 42 years, but autumn and winter bring 42 years of total darkness to the north pole.

What type of weather and climate changes does Uranus experience because of its extreme axis tilt? Astronomers don't know for sure. Although Uranus has been under telescopic scrutiny since its discovery by Herschel in 1781, we've only begun to see it clearly in the last two decades—the equivalent of one Uranian season. Thanks to the one-shot *Voyager 2* rendezvous in 1986 and, since the early 1990s, the keen eyes of the Hubble Space Telescope and large ground-based telescopes with adaptive optics, we've seen Uranus progress from summer in the southern hemisphere to autumn (winter to spring in the northern hemisphere). All we know so far is that around the time of summer for a particular hemisphere, Uranus tends to be rather quiet. As the equinox approaches, the atmosphere begins to show signs of activity. Spring weather across much of the continental United States is best described by the saying, "March comes in like a lion and goes out like a lamb." Astronomers are watching Uranus closely to see if there are lions and lambs in its seasonal weather.

....

Heidi Hammel (1960–)

Could chronic car sickness lead to a career in astronomy? It could if your name is Heidi Hammel. As a child, Hammel was prone to car sickness on the long trips she and her parents often took. As a distraction on night trips, she would lie back in her seat and gaze upward at the stars. Before long, she was able to recognize some of the major constellations.

The astronomy bug bit for good when Hammel took an astronomy course as a sophomore at the Massachusetts Institute of Technology (MIT). She graduated in 1982, following up six years later with a doctorate in physics and astronomy at the University of Hawaii.

In 1989, she was a member of the imaging team that analyzed the *Voyager 2* data from Neptune. Five years later, she led a ground-based team of scientists who studied Hubble Space Telescope images of the Great Comet Crash—the impact of Comet Shoemaker-Levy 9 with Jupiter.

Hammel continues her work as a planetary scientist, currently serving as senior research scientist and co-director of research at the Space Science Center in Boulder, Colorado. Much of her work centers on the ice giants Uranus and Neptune, which she studies with the Hubble Space Telescope and various ground-based instruments. She considers these often-overlooked planets the Rodney Dangerfields of planets because they don't get respect. Hammel is also involved with the planning of the James Webb Space Telescope, scheduled for launch in 2013. Designed to observe in infrared wavelengths, the Webb Telescope will help planetary astronomers like Hammel learn more about Uranus and Neptune.

....

A LOPSIDED MAGNETOSPHERE

Uranus is a topsy-turvy world in more ways than one. When we study the magnetospheres (magnetic fields) of the planets, in particular Jupiter and Saturn, we observe two basic truths—the magnetospheres are basically aligned with the planets' axes of rotation and are centered on the planets' cores. Uranus chooses to be different! Its magnetosphere is tilted almost 60 degrees to the axis of rotation. The field lines of that magnetosphere center on a spot one-third of the way out from Uranus's core. Because of this offset, the strength of Uranus's magnetic field varies from place to place on the planet. In general, the magnetic field strength near the cloud tops of Uranus is comparable to Earth's.

While the magnetic fields of the other planets seem to arise from electrical currents generated in their cores, this is obviously not the case with Uranus. Its source is as yet unknown, but is most likely generated by the movement of material in the mantle.

Like the magnetospheres of the other planets, the one surrounding Uranus interacts with the solar wind. On the side facing the Sun, Uranus's magnetic field extends outward until it encounters the incoming solar wind particles and is compressed. How far Uranus's magnetosphere extends sunward varies with the strength of the solar wind, but it is usually not more than about 460,000 kilometers (286,000 miles). The magnetotail on the other side stretches at least 10 million kilometers (6.2 million miles) into space. Because of the axis tilt of both Uranus's rotation and magnetic fields, the field lines in the magnetotail corkscrew into space.

The radiation belts associated with Uranus's magnetosphere are similar in strength to Saturn's but differ in composition. Primarily composed of electrons and protons, they lack heavier atoms, such as might be released from the surfaces of moons. These particles follow the magnetic field lines to Uranus, where they form faint auroras. Calculations show that the radiation present in Uranus's magnetosphere is intense enough to blacken any methane found on the surfaces of the inner moons and in the rings. This might explain why they appear so dark.

URANUS'S RINGS: A FALSE ALARM, THEN A SERENDIPITOUS DISCOVERY

Earlier in this chapter, we noted how the discovery of Uranus was accidental—the planet was found in 1781 by the astronomer William Herschel during an all-sky search for star clusters and nebulae. The discovery of the rings of Uranus nearly two centuries later would be yet another serendipitous event.

But first there was a false alarm. Eighteen years after discovering Uranus, Herschel reported sighting a ring around the planet. The observation turned out to be incorrect, and Herschel quickly retracted his announcement.

The rings were officially discovered in March 1977, when two teams of astronomers—one led by MIT's James L. Elliott on board the Kuiper Airborne Observatory, and the other at the Perth (Australia) Observatory—observed Uranus as the planet occulted (passed in front of) a bright star. Their plan was to use the fading of the star as it was covered by Uranus's disk to reveal telltale information about the planet and its atmosphere. To their surprise, the star flickered several times even before contact with the main planet. The pattern was repeated after the star reappeared from behind Uranus. The conclusion was that Uranus was surrounded by a system of faint, narrow rings. Analysis of the data from the two teams revealed the existence of nine separate rings encircling Uranus. They were named 6, 5, 4 (rings given numerical notations by the astronomers at Perth), alpha, beta, gamma, delta, epsilon, and eta. Saturn was no longer the only ringed planet in the solar system.

Two more rings, 1986 U1R and 1986 U2R (now known as lambda and zeta), were discovered by *Voyager 2* in 1986. Images of the two new rings plus the nine previously discovered were taken after *Voyager 2* passed Uranus. The rings were more easily seen when backlit by sunlight. Nearly 20 years passed before the Hubble Space Telescope imaged two faint outer rings, mu and nu, bringing the total to thirteen.

As is the case with Saturn's rings, the nomenclature of the Uranian ring system is a cosmic alphabet soup—made even more confusing because this ring alphabet uses Greek letters. In order, starting from the innermost and progressing outward, they are: zeta, 6, 5, 4, alpha, beta, eta, gamma, delta, lambda, epsilon, nu, and mu. They fall into three major groups. The first consists of nine narrow main rings, including 6, 5, 4, alpha, beta, eta, gamma, delta, and epsilon. The dusty inner rings (zeta and lambda) imaged by *Voyager 2* and the two remote rings (mu and nu) discovered by the Hubble complete the groups.

All of Uranus's rings occupy an area between 37,000 and 103,000 kilometers (23,000 and 64,000 miles) from the planet's center. Most are extremely thin and surprisingly dust-free, being composed of water ice intermixed with radiation-darkened chunks of organic material up to several meters in diameter. The rings are between 1 and 100 kilometers (0.5 to 62 miles) wide and perhaps a few hundred meters (yards) thick. Dust particles are found in a few rings, notably zeta, nu, and mu. This trio is wider than the others. Zeta and nu each span an area equal to the diameter of our Moon, while mu is five times wider still. Generally speaking, the Uranian ring system is a compromise between those of Jupiter and Saturn—dark like Jupiter's, but containing large particles like Saturn's.

By far, the most prominent of Uranus's rings is epsilon. It alone accounts for two-thirds of the brightness of the entire ring system. Epsilon, which combines dust-sized particles with chunks ranging between soccer balls and small cars in size, is similar to Saturn's F Ring. Like the F Ring, epsilon is extremely thin and is shepherded by a pair of moons, Cordelia on the inside and Ophelia along the outside edge.

Analysis of epsilon and its interaction with Cordelia and Ophelia indicates that this ring (and possibly the entire Uranian ring system) is no more than 600 million years old. A comparison of recent Hubble images of the rings with those taken two decades earlier by *Voyager 2* show dramatic changes. These dynamic rings likely are the remnants of a moon that broke apart near Uranus.

HOW DO YOU SAY IT?

How do you properly pronounce the name of the Sun's seventh planet? There are several options, but some lead to embarrassing consequences. Many a teacher, using the popular "your anus" pronunciation, has had to deal with the inevitable snickers when he or she proclaims, "You can see *your anus* with a telescope." Pronouncing it "urine us" has a similar outcome. Fortunately, the generally accepted pronunciation is "YOOR-un-nus." It sounds more official, and definitely cuts down on the giggle factor.

SHAKESPEAREAN MOONS (WITH AN ASSIST FROM ALEXANDER POPE)

The first two of Uranus's 27 known moons were initially seen by William Herschel in 1787, about a half dozen years after he discovered Uranus. Two more fell to the watchful eye of another Englishman, William Lassell, in 1851. These four moons remained nameless until the following year when William Herschel's son John suggested names taken from English literature. The moons discovered by his father were christened Oberon and Titania, after characters in Shakespeare's *A Midsummer Night's Dream*. Lassell's moons were dubbed Umbriel and Ariel, the names taken from Alexander Pope's *The Rape of the Lock*. The younger Herschel's idea wasn't popular with traditionalists who preferred names from Greek and Roman mythology, but it survived and continues in use today. In fact, when a fifth Uranian moon was discovered on a photograph taken by Gerard Kuiper in 1948, it was named Miranda, after a character in Shakespeare's *The Tempest*.

The satellites of Uranus can be loosely placed in three groups. Close to the planet are 13 small and very dark moons, of which *Voyager 2* discovered 11 during its 1986 flyby. Most are associated with the planet's rings. Next come the original five satellites discovered between 1787 and 1948. None of these are giants like Jupiter's Galilean moons or Saturn's Titan. The largest, Titania, is less than half the diameter of our Moon. These five are distinguished from the moons of Jupiter and Saturn in that they have greater overall densities and may be a half-and-half blend of rock and water ice. The remaining moons all orbit at great distances from Uranus. Discovered between 1997 and 2003 by large ground-based telescopes and the Hubble Space Telescope, most are tiny and have retrograde orbits. Like the outer moons of Jupiter and Saturn, they may be captured bodies.

THE INNER MOONS: KEEPERS OF THE RINGS

When the rings of Uranus were discovered in 1977 and found to be extremely thin, astronomers speculated the existence of a collection of tiny shepherd moons like the ones that shape Saturn's rings. *Voyager 2* proved them right, finding 10 new moons (plus an 11th, later discovered during in-depth analysis of a *Voyager 2* photograph). In 2003, two more moons were uncovered by the Hubble Space Telescope; the 10-kilometer (6-mile)-wide Mab, outermost of the pair, is associated with one of Uranus's faint, dusty rings.

Two of the *Voyager 2* moons were found to shepherd Uranus's epsilon ring. These innermost of Uranus's satellites, Cordelia (cor-DEE-lee-uh) and Ophelia (oh-FEE-lee-uh), are between 25 and 30 kilometers (15.5 and 18.6 miles) across and travel around Uranus in eight and nine hours, respectively.

Of Uranus's moons, the two found by Hubble are the smallest, with estimated diameters of 10 kilometers (6 miles). The remaining 11 moons range in size from 20 to 160 kilometers (12 to 100 miles). Largest is heavily cratered Puck. Its dark surface is typical of the group, and may be caused by the blackening of surface methane as it interacts with the charged particles found in Uranus's magnetosphere.

The orbits of these inner satellites fill a zone just 48,000 kilometers (29,825 miles) across, starting 49,750 kilometers (30,900 miles) from the center of Uranus. They move swiftly with periods between 8 and 22 hours. Thirteen moons racing around in an area no wider than four Earths is a recipe for disaster. As they zip past one another, gravitational tugs could alter their orbits, leading to catastrophe at some point in the future. Indeed, it's possible that Uranus's rings were created by earlier collisions.

MIRANDA, A REASSEMBLED MOON?

Pronounced: "mih-RAN-dah"
Diameter: 470 km (290 mi)
Mean Distance from Uranus: 129,872 km (80,700 mi)
Orbital Period: 1.41 days
Period of Rotation: 1.41 days (synchronous)
Average Density: 1.2 g/cm^3
Surface Gravity, Compared to Earth: 0.8%

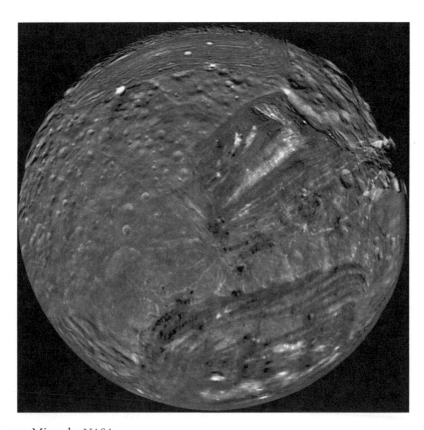

Figure 8.2 Miranda. NASA.

Smallest of Uranus's five major moons (its equatorial diameter is just one-seventh that of our Moon), Miranda was discovered on a photographic plate taken by Gerard Kuiper in 1948. Too puny to merit much attention, Miarnda wasn't a high-priority target for the *Voyager 2* flyby. Fortunately, it was imaged, as were the other four and the 11 tiny inner moons never before seen. The photographs of Miranda were nothing short of spectacular, revealing an amazing array of surface features for such a small body. Geologically, Miranda ranks as one of the solar system's most fascinating moons.

Particularly noteworthy are the chevron-shaped fault zones, highlighted by cliffs that overlook a 20-kilometer (12-mile) drop—12 times the depth of the Grand Canyon. Miranda's jumbled terrain has led some astronomers to speculate that this moon was shattered by one or more past cosmic collisions, each time gravitationally reassembling itself like a jigsaw puzzle whose pieces don't quite interlock. It's also possible that Miranda experienced a series of upwellings in its early history, with hot material from the interior rising to the surface and quickly hardening.

Miranda circles Uranus in a synchronous orbit once every 1.41 days. Surface temperatures on this frigid moon hover at −187°C (−305°F). Having an average density of 1.2 g/cm^3, Miranda, like the other major moons of Uranus, is likely a 50-50 mix of rock and ices.

ARIEL

Pronounced: "AIR-ee-al"
Diameter: 1,158 km (720 mi)
Mean Distance from Uranus: 190,945 km (118,655 mi)
Orbital Period: 2.52 days
Period of Rotation: 2.52 days (synchronous)
Average Density: 1.6 g/cm^3
Surface Gravity, Compared to Earth: 2.7%

Possessing the most reflective and possibly youngest surface of Uranus's five major moons, Ariel is characterized by a cratered surface broken by long intersecting fault valleys. The floors of these valleys appear smooth, as if eroded by a fluid, possibly ammonia or methane. Another theory is that, like some of Saturn's moons, the surface solidified first, and then as the interior froze and expanded, the surface fractured. Subsurface material oozed to the surface, filling in the cracks.

Few of Ariel's craters exceed 50 kilometers (30 miles) in diameter, and those larger in size have been resurfaced. Like Miranda, this rock/ice moon is locked in a synchronous orbit around Uranus. Ariel was discovered by English astronomer William Lassell in 1851.

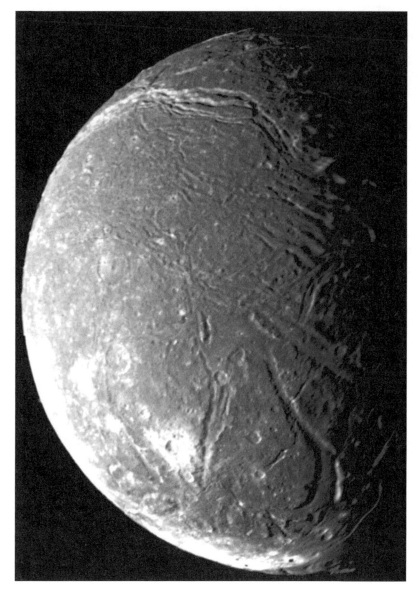

Figure 8.3 Ariel. NASA.

UMBRIEL

Pronounced: "UM-bree-el"
Diameter: 1,170 km (727 mi)
Mean Distance from Uranus: 265,998 km (165,290 mi)
Orbital Period: 4.14 days
Period of Rotation: 4.14 days (synchronous)
Average Density: 1.5 g/cm^3
Surface Gravity, Compared to Earth: 2.4%

Figure 8.4 Umbriel. NASA.

Umbriel was also discovered by Lassell in 1851. A move from Ariel to Umbriel takes us from Uranus's brightest moon to its darkest. Its surface is older and uniformly cratered. Other than that, the two moons are similar in size, density, and composition. Umbriel also keeps one side facing Uranus during its orbital dance.

A *Voyager 2* photograph shows a bright spot near Umbriel's edge. This bright 130-kilometer (80-mile)-wide ring, called "Wunda" but nicknamed the "fluorescent cheerio," may be the floor of a crater. Astronomers aren't quite sure why it's so much brighter than the rest of Umbriel.

TITANIA

Pronounced: "ti-TAY-nee-uh"
Diameter: 1,578 km (980 mi)
Mean Distance from Uranus: 436,298 km (271,115 mi)
Orbital Period: 8.71 days
Period of Rotation: 8.71 days (synchronous)
Average Density: 1.7 g/cm^3
Surface Gravity, Compared to Earth: 3.9%

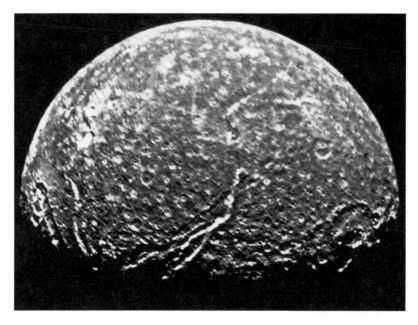

Figure 8.5 Titania. AP Photo/NASA.

Largest of Uranus's moons by a very slight margin over Oberon, Titania is still less than half the size of our Moon. A scaled-up version of Ariel, Titania also mixes cratered terrain with long interconnected valleys hundreds of kilometers long. One of its major surface features is the multiring basin Gertrude. Over 300 kilometers (186 miles) across, it could hold within its boundaries the entire state of Maine. The Messina Chasmata is a valley that traverses 1,600 kilometers (1,000 miles) of Titania's surface and would easily dwarf the Grand Canyon. Titania was discovered by William Herschel in 1787.

OBERON

Pronounced: "OH-buh-ron"
Diameter: 1,523 km (946 mi)
Mean Distance from Uranus: 583,519 km (362,600 mi)
Orbital Period: 13.46 days
Period of Rotation: 13.46 days (synchronous)
Average Density: 1.6 g/cm^3
Surface Gravity, Compared to Earth: 3.5%

Oberon is comparable in size to Titania, but more similar in appearance to Umbriel. Like Umbriel, Oberon is characterized by an ancient, heavily cratered surface. The floors of some of Oberon's craters are covered by dark material, indicating an upwelling of some unknown subsurface substance.

Figure 8.6 Oberon. NASA.

Typical of these is the 296-kilometer (184-mile)-wide crater Hamlet, whose dark floor reminds the observer of some of our Moon's walled plains, like Plato. Other craters are surrounded by bright rays of ejected material similar in appearance to those seen on Jupiter's moon Callisto. The *Voyager 2* photograph reveals a bump on Oberon's edge. This bump is the peak of a mountain that appears to be peeking over the edge of Oberon. Astronomers guess that it's at least 6.4 kilometers (4 miles) high. Like Titania, Oberon was discovered by William Herschel in 1787.

THE OUTER MOONS

Like Jupiter and Saturn, Uranus is attended by a family of distant moons, most of which are small and travel in highly inclined, retrograde orbits. These, too, were likely captured bodies. All were discovered since 1997 by ground-based telescopes, mostly by a team of astronomers that included Brett Gladman, John Kavelaars, and Matthew Holman. These moons orbit

Uranus at distances ranging from 4,276,000 to 20,901,000 kilometers (2,657,000 to 12,988,000 miles) in periods of nine months to eight years.

The largest of Uranus's outer moons is Sycorax (SIK-or-aks), with a diameter of about 150 kilometers (93 miles). Margaret, a city-sized moon discovered in 2003, is a misfit in the group. It orbits Uranus in the same direct motion as the main moons. Its orbit is also much less inclined. The most distant of Uranus's known moons is Ferdinand, estimated to be about 20 kilometers (12.5 miles) in diameter. Ferdinand is one of the smallest Uranian satellites yet discovered. Many more are likely to be discovered as the optical capabilities of ground-based telescopes improve.

URANUS'S MOONS: THE MOST POPULAR FUTURE TOURIST SITES

While we citizens of the early twenty-first century daydream of vacations in faraway places, perhaps a safari in Africa or a cruise to Alaska's glaciers, our descendants will set their sights on truly out-of-this-world locations. In Chapters 5 and 7, we reviewed future tourist sights on the moons of Jupiter and Saturn. For those who seek real out-of-the-way destinations, here are some suggestions, compliments of Travel Uranus.

1. Miranda. Your great-grandfather might have boasted about parasailing off the cliff wall at El Capitan in Yosemite National Park or parachuting out of an airplane three miles above the ground. We'd bet he would have gladly traded those experiences for a chance to free fall off one of the cliffs on Uranus's amazing moon Miranda. After all, Miranda offers a drop *20 times greater* than El Capitan and *7 times higher* than a high-altitude parachute jump. While parasails and parachutes won't work in Miranda's airless environment, the moon's low gravity (it's just 1/125 Earth's) allows you to fall at a relaxed rate, as if you were parachuting or parasailing. At the end of your 12-minute free fall, rockets on your backpack allow for a safe and comfortable landing. Finish your adventure with a relaxing visit to our lodge, where guests are offered an eerily spectacular view of blue-green Uranus.

2. Titania. If you like to mix the excitement of scientific discovery with awe-inspiring scenery, you can always join a geological dig in the Grand Canyon on Earth or on Mars's Vallis Marineris. Better yet, join us on an excursion to the largest Uranian moon, Titania. With a genuine geologist (or should we say Titaniologist?) as your guide, you'll explore the magnificent Messina Chasmata. As you collect samples from the valley floor, you'll marvel at the majesty of walls that rise to heights of 10 kilometers. If you thought walking across a frozen pond in winter was unique, wait until you step on ice so cold that it's as hard as steel!

SEE FOR YOURSELF

Jupiter and Saturn are easy targets for the novice skygazer because they are bright and readily found in the evening sky. This is not the case with Uranus. Barely visible to the unaided eye on a clear, dark night, this planet requires knowledge of the constellations and the ability to navigate the night sky with a finder chart.

Uranus is easily seen with binoculars, but you won't see much more than a blue "star." Still, it's fascinating to realize that the light striking your eyes left Uranus 2.5 hours earlier! To see the planet's tiny disk, equal to the apparent size of a golf ball placed two kilometers (1.3 miles) away, you'll need a telescope and a power of at least 100X. Such a magnification is within the reach of most common backyard scopes.

To see detail on Uranus's disk or to observe its moons requires a large telescope, perfectly steady atmospheric conditions, and a skilled eye. Only the most experienced amateur astronomers, many using CCD photography, capture such detail.

Should you seriously want to see Uranus with your own eyes, but lack the experience or equipment necessary to do the job, why not seek the help of a local amateur astronomer? If you don't know anybody first-hand, contact a local high school or college where astronomy classes are taught. A great way to get in touch with an amateur astronomer is by contacting a local astronomy club. To find one near you, log on to www.astronomyclubs.com. Your nearby astronomy club may have public outreach programs, and many of its members are only glad to help beginners discover the night sky. As you gaze at that tiny blue disk in the eyepiece field, think how much its appearance intrigued William Herschel when he stumbled upon it on that March evening in 1781.

WEB SITES

http://www.nineplanets.org/uranus.html.
    A Nine (8) Planets Web site look at Uranus.
http://www.astronomycast.com/astronomy/episode-62-uranus.
    The script to "Astronomycast's" episode on the planet Uranus. Includes links to the podcast and useful Web sites relating to Uranus.
http://sse.jpl.nasa.gov/planets/profile.cfm?Object=Uranus.
    A NASA Web site containing important facts about Uranus.

9

Neptune, the Planet Discovered on Paper

August 25, 1989: *Voyager 2* has arrived at the final target of the Grand Tour, the planet Neptune. This incredible machine defied the odds and survived the perilous trek to the four outer planets. It will send remarkable images of Neptune and its largest moon, Triton, before continuing onward out of the solar system. As one NASA official notes, it is the "final movement in the *Voyager* symphony."

On Earth, some 4.5 billion kilometers (2.8 billion miles) away, Michelle peers through her telescope at Neptune—a mere speck identifiable only with the help of a star chart. Michelle is now a high school honors student who plans to major in astronomy when she enters college. She gazes in awed silence, knowing that a piece of machinery launched from Earth a dozen years earlier is now there. "Thanks and good luck *Voyager 2*," she whispers before packing up her telescope and returning indoors.

NEPTUNE'S DISCOVERY: A PLANET FOUND ON PAPER

It took a diligent observer (William Herschel), using a telescope whose light-gathering mirror was 15 centimeters (6 inches) across to discover Uranus. What sort of equipment would astronomers need to spot a planet 1.6 billion kilometers (1 billion miles) farther out in the solar system? You might assume that a bigger telescope was used, or perhaps a photographic

Neptune Data

Period of Revolution	164.8 years
Period of Rotation	$16^h 7^m$
Axis Tilt	28.3°
Equatorial Diameter	49,538 kilometers (30,776 miles)
Mass (Earth = 1)	17.1
Surface Gravity* (Earth = 1)	1.1
Density (water = 1.0 g/cm³)	1.64
Number of Moons	13
Mean Distance from Sun	30.1 AU (4,498,253,000 kilometers [2,795,085,000 miles])

* Since none of the Jovian planets has a solid surface, gravity is calculated at the visible cloud tops.

plate. The fact is, Neptune was essentially discovered with two very common items—paper and pencil.

The account of the discovery of Neptune in 1846 by two mathematicians who predicted its existence and location before it fell to human eyes is as steeped in mystery as the planet itself. For over a century and a half, the

Figure 9.1 Neptune. AP Photo/NASA.

traditional story was that a young English mathematician, John Couch Adams, had toiled for several years in an effort to find a planet astronomers hypothesized was causing unusual behavior in the orbital motion of Uranus. When Adams forwarded his calculations to George Airy, the English Astronomer Royal, he was largely ignored. Meanwhile, on the other side of the English Channel, French mathematician Urbain Jean Le Verrier began attacking the same problem. Upon coming up with the location of the phantom planet he, too, was unable to capture the attention of his country's astronomers. Undaunted, he mailed his calculations to the German astronomer Johannes Galle, who found the new planet on the first night he looked.

What ensued was an ugly confrontation between England and France over which mathematician deserved credit for the find. In the end, Adams and Le Verrier were given equal credit. How sad, the tale goes, that poor Adams would have to share the honors with Le Verrier, when he had correctly predicted Neptune's location a year earlier! In recent years, the discovery of long-lost documents indicates that Adams's calculations may not have been as accurate as originally claimed, and that it was Le Verrier who was robbed.

The saga of the planet discovered on paper goes back over 60 years earlier. During the first four decades after its discovery in 1781, Uranus had been orbiting the Sun faster than calculations predicted. In 1822, however, the planet began to slow down. Had a comet struck Uranus? Not likely. Astronomers and mathematicians, relying on Newton's laws of gravity, began considering the idea that a more distant and as yet unseen planet might be the culprit, tugging at Uranus as the two passed in their orbits. But because there were still uncertainties in Uranus's orbit and the mathematics of finding the unknown planet (if one even existed) were daunting, no one seriously attacked the problem.

Enter Adams. A brilliant young mathematician, he was a student at the prestigious Cambridge University in 1841 when he first heard of the problem with Uranus's orbit. At that moment, he resolved to locate the mischievous unseen planet whose gravity was bullying Uranus. After his graduation in 1843, Adams began working on the problem in earnest.

In the summer of 1845, the French scientist Jean Dominique Francois Arago suggested to Urbain Le Verrier that he tackle the Uranus problem. Unlike Adams, who was young and relatively unknown, Le Verrier was an honored and established mathematician. Within a year, Le Verrier had worked out the location of the new planet and presented his findings to the Paris Academy of Science in June of 1846. Amazingly, he was unable to find an astronomer who would follow up with a visual search. An updated paper on the new planet, presented a few months later, again failed to elicit response.

When word of Le Verrier's work reached George Airy, the Astronomer Royal realized that both Adams and Le Verrier were pointing to the same general area of sky as the location for the new planet. This prompted Airy to ask James Challis, director of the Cambridge Observatory, to begin looking. Challis began sweeping the skies with his telescope in August and September.

Le Verrier meanwhile decided that if no one in France would look for the planet, perhaps someone elsewhere would help. He turned to the German astronomer Johann Gottfried Galle, writing him a letter with details about the planet's predicted location. Galle received the correspondence on September 23, 1846. That very night, he commenced a search, using the 23-centimeter (9-inch) refractor at the Berlin Observatory. Assisted by Heinrich D'Arrest, Galle began a star-by-star search. Within an hour, they came across an 8th magnitude star not shown on the charts. The next night, they revisited the object to see if it had moved against the background stars. Not only had it moved, but it showed a tiny planetary disk under high magnification. The new planet had been discovered!

In those days, any major cosmic discovery was a source of pride for the astronomer's home country. While a German had visually identified the new planet, it was the brilliant work of a man from France—Urbain Le Verrier—that brought it to light. Around the world, scientists hailed the discovery of the new planet as a triumph of mathematics and a validation of Newton's laws of gravity.

We can only imagine how Airy, Challis, and Adams felt when they learned about the discovery of the new planet. When Challis checked his records, he discovered that he had seen the new planet on two separate occasions but had failed to recognize it as such. A claim of "you found it first, but we predicted it first" arose from the English. The French were not pleased. The argument took on the proportions of an international crisis.

Based on evidence from the recovered documents, we now know that Adams's calculations were not as accurate as the English claimed—in fact, he may have been as much as 20 degrees (about 40 Moon diameters) off. While Adams was indeed a brilliant mathematician, his contribution to the discovery of Neptune was quite likely overstated. Perhaps, over 150 years after the fact, we should give Le Verrier the individual honor he deserves.

Naming the new planet was no less controversial than the circumstances of its discovery. Galle suggested the name Janus, while Challis offered Oceanus. Le Verrier at first recommended the name Neptune, but later reconsidered, asking that the planet bear his name. To legitimize the naming of a planet for its discoverer, Le Verrier and his fellow countrymen resorted to referring to Uranus as "Herschel." The name "Le Verrier" for the new planet was about as popular outside of France as the aliases "George's Planet" and "Herschel" had been for Uranus! In the end, the name Neptune was adopted.

....

Urbain Le Verrier (1811–1877)

Few individuals were better equipped to calculate the position of an unknown planet than the brilliant French mathematician Urbain Jean Le Verrier. Born in St. Lo, France, and educated at the *École Polytechnique*, Le Verrier spent most of his working life at the Paris Observatory. It was here that the Observatory's director, Jean Arago, suggested that Le Verrier use his mathematical skills to help in the search for the mystery planet believed to be gravitationally altering the orbit of Uranus.

After the discovery of Neptune, Le Verrier used the motions of Mercury to predict the existence of another planet, Vulcan, even closer to the Sun. The planet was never found, and Mercury's supposed orbital disturbances were later explained by Einstein's theory of general relativity. In his later years, Le Verrier refined and established the orbits of the eight planets. He also served as director of the Paris Observatory.

....

....

John Couch Adams (1819–1892)

John Couch Adams is best remembered (perhaps incorrectly, as recently recovered records indicate) as the unfortunate young English mathematician who made the first accurate calculations of the location of Neptune, but failed to get the proper attention of his country's astronomers. Whatever the truth, Adams was a gifted individual who showed his mathematical prowess at an early age.

To his credit, Adams never expressed any bitterness to being Neptune's "runner-up" discoverer, even forming a lifelong friendship with Le Verrier. After the discovery of Neptune, Adams continued his contributions to mathematical astronomy, proving that the great Leonid meteor shower is associated with a comet. He ultimately became the director of the Cambridge Observatory, a position he held until his death.

....

....

A Near Discovery for Galileo

The saga of Neptune's paper-and-pencil discovery might never have happened had an Italian astronomer correctly identified a "star" near Jupiter more than two centuries earlier. The star appears in sketches of Jupiter and its moons that he drew in 1612 and 1613. The name of the astronomer who missed an opportunity to discover Neptune—Galileo Galilei!

....

GETTING TO KNOW A NEW PLANET

Uranus is so far away that it's hard to observe through earthbound telescopes. Neptune, half again as distant and 10 times fainter, is a near impossibility. Fortunately, the detection of a large moon, Triton, by the English astronomer William Lassell just 17 days after Neptune was discovered gave astronomers the key to mathematically unlock Neptune's mass. Lassell's discovery of a Neptunian moon might have given his countrymen some consolation in the Neptune discovery controversy. While Lassell was yet another English astronomer who had failed to look for Neptune, he had a valid excuse. At the time he received a request to participate in the search, he was bedridden with a badly sprained ankle.

Rough measurements of Neptune's disk, equal in apparent size to a golf ball 3.7 kilometers (2.3 miles) away, revealed that the new planet might be a twin in diameter to Uranus. Its orbital period was established at 164.8 years, twice that of Uranus.

Calculations showed that Neptune, like Jupiter, Saturn, and Uranus, is a low-density planet. It came as no surprise, therefore, when the spectroscope showed that Neptune's atmosphere is dominated by hydrogen with traces of methane. There was even evidence that Neptune's large moon Triton might hold an extremely tenuous atmosphere. In 1949, Gerard Kuiper photographed a second moon, dubbed Nereid. Triton circles Neptune in a retrograde orbit, which leads some astronomers to suspect it is a captured object.

During its 1989 Neptune encounter, *Voyager 2* confirmed the existence of Neptune's rings, whose presence had been hinted at during a series of occultations of stars several years earlier. What astronomers had thought were "ring arcs" turned out to be denser parts of the main ring. Six new moons were discovered. One, Proteus, is even bigger than Nereid, but too dark and close to Neptune to be readily seen through Earth-based telescopes. *Voyager* imaged and analyzed Neptune's surprisingly active weather patterns and mapped its magnetic field.

Like the encounter with Uranus, the *Voyager 2* flyby of Neptune was a "one-shot deal." There are no immediate plans to send robotic probes to either world in the near future. Adaptive optics on Earth-based telescopes, augmented by images taken by the Hubble Space Telescope, have kept astronomers in touch with Neptune. Thanks to these modern technologies, astronomers have added five new moons and monitored weather patterns in Neptune's rapidly changing atmosphere.

It is interesting to note that from the time of its discovery on September 23, 1846, until the time this book goes to press in 2009, Neptune will not have completed a full revolution around the Sun. The first Neptunian year since its discovery will occur on July 12, 2011.

Until its demotion to "dwarf planet" status by the International Astronomical Union in the summer of 2006, Pluto was considered the Sun's most distant planet (except for a 30-year period when Pluto's highly eccentric orbit cuts inside Neptune's). Neptune now holds that distinction. However, some theorists believe that Neptune might have been closer to the Sun than Uranus at the time the solar system formed. As noted in the last chapter, astronomers looking at the compositions of Uranus and Neptune believe that these planets couldn't have been created in their current locations. According to the core-accretion theory, there wasn't enough material in this part of the solar nebula to accrete, or merge together, into such large planets. It's speculated that both Uranus and Neptune formed close to Jupiter and Saturn in an area rich with planet-building materials. Gravitational interactions with Jupiter and Saturn and intermingling icy planetesimals pushed Uranus and Neptune outward to their present orbits. Neptune may have been the closer of the two to the Sun before being flung past Uranus.

A TWIN TO URANUS?

Although each possesses its share of unique characteristics, Neptune and Uranus are about as close to planetary "twins" as you can get in the solar system. Both are huge, having diameters four times Earth's. Their atmospheres are comprised mainly of hydrogen and helium, with a small quantity of methane and ammonia. The methane absorbs red light, giving the planets their distinct bluish color. Even the rotation periods of Uranus and Neptune are similar, differing by only about an hour.

Uranus and Neptune appear to occupy a niche midway between the gas giants Jupiter and Saturn and the four terrestrial planets. Neptune is 17 times more massive than Earth and one-nineteenth as massive as Jupiter. While Jupiter and Saturn seem to be almost totally constructed of hydrogen and helium, Uranus and Neptune possess these gases, plus higher quantities of the "ices" water, methane, and ammonia. For this reason, astronomers have recently begun to refer to Uranus and Neptune as "ice giants."

Analysis of Neptune's atmosphere shows that it's comprised of 79 percent hydrogen, 18 percent helium, and 3 percent methane. Our best guesses of what lies beneath the clouds come from computer-generated models based on *Voyager 2* data about the planet's size, atmospheric composition, and overall density. According to the most popular model, the gaseous hydrogen-helium-methane atmosphere covers a hot, soupy, highly pressurized mantle of ices, primarily water, methane, and ammonia. Near the planet's center is a core of rock and ice. This model is almost identical to the one we have for Uranus.

What distinguishes Neptune from its "twin brother" Uranus? An obvious difference is in the tilt of their axes of rotation. We've already seen that Uranus has an axis tipped so dramatically that it literally rolls on its side as it orbits the Sun. Neptune brings us back to normalcy. Its axis tilt of 28.3° is similar to Earth's. The two ice giants also differ in mass and density, Neptune being slightly smaller than Uranus, but denser and more massive.

A close look at the weather that exists on the two ice giants will reveal another major difference. When *Voyager 2* began its approach to Neptune, astronomers had low expectations. After all, Uranus had proved to be relatively bland and featureless, releasing only a tiny amount of energy beyond what it receives from the Sun. Neptune, 1.6 billion kilometers (1 billion miles) further from the Sun could hardly be expected to be a hotbed of meteorological activity. But, the planetary scientists were in for a surprise.

Even before *Voyager* arrived at Neptune, long-range photographs picked up a huge, dark, oval-shaped spot in Neptune's atmosphere. Large enough to hold the Earth within its boundaries, this 6,600 by 14,000 kilometers (4,100 by 8075 miles) storm proved to be a Neptunian version of Jupiter's Great Red Spot. Besides the Great Dark Spot, large, rapidly moving white clouds nicknamed "Scooters" could be seen.

Despite some differences, the overall makeup of Neptune's clouds mirrors what astronomers have found on Uranus. Below a hydrocarbon haze, layers of clouds comprised of methane, ammonia, hydrogen sulfide, ammonium sulfide, and water exist at various levels.

If meteorologists can refer to Chicago as the "Windy City," planetologists are within their rights to nickname Neptune the "Windy Planet." High-altitude winds near Neptune's equator were measured by *Voyager 2* at speeds as high as 2,160 kilometers (1,340 miles) per hour. On Earth, such a wind would cross the continental United States in a little over two hours. As with the other Jovian planets, Neptune's weather patterns are confined to distinct bands running parallel to the equator.

The high winds of Neptune aren't as ferocious as you might imagine. On Earth, a breeze blowing at a 32 kilometer (20 mile) per hour clip would barely ruffle your clothing. However, floodwaters flowing at this speed would sweep away automobiles. The difference is that water is much denser than air and would exert a greater "push." Neptune's supersonic winds occur high in the planet's atmosphere. The gases there are far less dense than the air here on the Earth's surface. It's possible that Neptune's high-altitude high-speed winds would hardly be able to spin the blades of an Earthly wind turbine.

Earth's weather is generated by the Sun. On Neptune, as is the case with the other Jovian planets, the Sun is too remote to drive weather patterns. The heat source for Neptune's weather comes from within. Instruments on board *Voyager 2* showed that Neptune releases into space over twice the amount of heat energy it receives from the Sun. This is a dramatic departure from the situation with Uranus, where there is little appreciable difference between the energy received and energy returned to space. What is the source of Neptune's internal heat? Could it be the radioactive decay of elements in Neptune's core? Is the planet still contracting? Is there some exotic source presently not understood? Astronomers don't know.

The climate on Neptune is another astronomical mystery. Seasons on Uranus last more than 20 years; those on Neptune are twice as long. The *Voyager 2* flyby of Neptune in 1989 gave planetary scientists their first detailed weather report of this planet. Hubble and large ground-based scopes continue to provide updates. One such update, made just six years after the *Voyager 2* flyby, showed that the Great Dark Spot was gone! A Neptunian storm that had been compared to Jupiter's long-lived Great Red Spot had disappeared within a half dozen years. Still, two decades of Neptunian weather is comparable to about six weeks of sampling Earth's weather, and six weeks out of a year says little about climate. As astronomers obtain daily weather reports from Neptune during the decades ahead, they'll be able to piece together a picture of the big planet's climate.

Diamonds in the Sky?

In 1999, scientists at the University of California at Berkeley decided to simulate the conditions found in Neptune's interior approximately one-third of the way from the cloud tops to the core. When they subjected a sample of methane (a compound comprised of carbon and hydrogen) to the temperature and pressure believed to exist in that zone, an amazing thing happened. The ammonia dissipated, leaving behind carbon that was transformed into tiny diamonds! Could there be a "rainfall" of fine diamond dust in Neptune's interior, and might friction from this falling dust contribute to Neptune's internal heat?

ANOTHER LOPSIDED MAGNETOSPHERE

Neptune is a twin to Uranus in more ways than just size and composition. The planet's magnetosphere, like the one associated with Uranus, is lopsided and centered on an area offset from the planet's core. In Neptune's case, the magnetic and rotational axes differ by 47°. Neptune's magnetic field seems to be centered in a zone of electrically conducting fluids about midway between the planet's core and the cloud tops.

The magnetic fields of the outer planets release radio waves. Jupiter's are strong enough to be detected here on Earth. Those of the other three Jovian planets, too weak to reach Earth, were detected by instruments aboard *Voyager 2*. Since the magnetic fields of the Jovian planets spin in close agreement with their cores, astronomers are able to pinpoint their true rates of rotation. Analysis of Neptune's magnetic field led to the 16.1-hour period of rotation given in modern-day textbooks. Neptune's magnetosphere, shaped by the solar wind, extends about 625,000 kilometers (388,375 miles) toward the Sun. The magnetotail stretches at least three times that distance in the opposite direction.

THE FOURTH RINGED PLANET

After the discovery of a system of rings around Uranus when the planet occulted a star, astronomers paid close attention to occultations of stars by Neptune. Such events would be even rarer, because Neptune's disk is only about half the size of Uranus's. Still, astronomers were able to observe a number of these events in the early to mid-1980s. The results were inconclusive. Unlike the Uranus occultations, where the pattern of star dimming was symmetrical on both sides of the planet, those exhibited by Neptune were not. Were the blinks caused by a host of tiny moons or by incomplete ring arcs? Science would have to wait for *Voyager 2* to provide answers.

Voyager 2 discovered a total of five faint rings surrounding Neptune. The rings were complete but not uniform, characterized by thick clumps of material apparently confined by shepherding moons. Neptune's rings appear to be similar to those that surround Jupiter—extremely dark, faint, and dust-laden.

The five rings were named in honor of astronomers and mathematicians who played roles in the discovery of Neptune. Closest to Neptune and progressing outward, they are Galle, Le Verrier, Lassell, Arago, and Adams. You recall that Le Verrier and Adams independently calculated Neptune's hidden location from its gravitational tug on Uranus, and that Galle was the astronomer who used Le Verrier's predictions to confirm Neptune's existence telescopically. Jean Arago, in his role as director of the Paris Observatory, was the one who encouraged Le Verrier to look for the unseen planet. William Lassell was an English astronomer who discovered Neptune's largest moon, Triton, just weeks after the planet was found. The five rings occupy a zone that begins 41,000 kilometers (25,500 miles) from Neptune and extends outward another 22,000 kilometers (13,000 miles).

The Adams Ring is the most complex, and is comprised of five dense areas, or ring arcs, named Fraternity, Equality 1, Equality 2, Liberty, and Courage. Astronomers were baffled by these arcs, because material concentrated in one part of a planetary ring should naturally spread throughout the entire ring. The arcs in the Adams Ring likely result from the gravitational influence of the nearby moon Galatea. Astronomers studying the Adams Ring with the Hubble Space Telescope and enhanced optics of ground-based instruments have detected a dramatic breakdown in some of the ring arcs since the *Voyager 2* flyby. Perhaps they are temporary after all.

NEPTUNE'S MOONS, SURVIVORS OF CHAOS?

Neptune takes the "all or nothing" scenario of Jupiter's moons to an extreme. Only one satellite, Triton, is large enough to be spherical. The rest are so small that their combined mass is less than 1 percent of Triton's. Neptune's 13 known moons are a fraction of the number circling the other outer planets. Because of Neptune's great distance, it may possess dozens of moons between one and 30 kilometers (0.5 and 18.5 miles) in diameter that wait to be discovered. On the other hand, Neptune may be truly "moon-poor," as Jovian planets go. Astronomers speculate that Triton, which may be a captured moon, collided with and destroyed a number of moons when it entered the Neptunian system.

Triton was found within weeks of Neptune's discovery. A century passed before a second moon, Nereid, was discovered photographically. The pair and all of Neptune's moons discovered subsequently were named for lesser sea gods and nymphs in mythology. Six moons were imaged by *Voyager 2* during its 1989 flyby. The largest, Proteus, has a diameter slightly greater than Nereid's. The rest of *Voyager*'s finds are all small and dark, and orbit in and around Neptune's rings. A final group of satellites located extremely far from Neptune were picked up in 2002–2003 by Earth-based telescopes, with enhanced adaptive optics.

NEPTUNE'S INNER MOONS

Neptune is orbited by six inner moons, all discovered on images taken by the *Voyager 2* probe in 1989. From Neptune outward, they are Naiad (NYE-uhd), Thalassa (thuh-LASS-uh), Despina (dess-PEE-nuh), Galatea (GAL-uh-TEE-uh), Larissa (luh-RISS-uh), and Proteus (PRO-tee-us). Larissa was actually detected in 1981 during an occultation by Neptune of a distant star. At the time, astronomers attributed the star's temporary dip in brightness to a possible partial ring surrounding Neptune. The four innermost of the group orbit within Neptune's ring system, and two, Despina and Galatea, have shown direct gravitational influence on the rings. With the exception of Proteus, none of these moons is larger in diameter than 220 kilometers (137 miles). All but Proteus circle Neptune faster than the planet rotates, having periods between 7 and 13 hours.

PROTEUS

Pronounced: "PRO-tee-us"
Diameter: irregular; average 400 kilometers (250 miles)
Mean Distance from Neptune: 117,650 kilometers (73,105 miles)
Orbital Period: 1.12 days
Period of Rotation: 1.12 days (synchronous)
Average Density: 1.3 g/cm^3 (estimated)
Surface Gravity, Compared to Earth: 0.6%

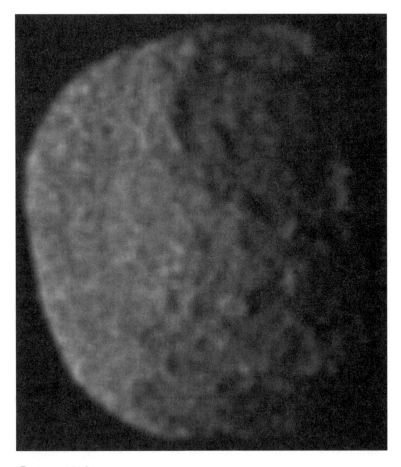

Figure 9.2 Proteus. NASA.

Proteus is the largest of Neptune's six inner moons. Second in size among the Sea God's moons, it eluded detection by Earth-bound telescopes primarily because of its darkness and close proximity to the planet. Neptune's glare may have made Proteus all but impossible to see from Earth, but it wasn't enough to keep this moon from the watchful eye of *Voyager 2.* Three times closer to Neptune than our Moon is from Earth, Proteus zips around its home planet in a little over one Earth day.

Like Saturn's moon Phoebe, Proteus is extremely dark ("as dark as soot," says a NASA description), reflecting a mere 6 percent of the sunlight that strikes it. The *Voyager 2* image of Proteus shows a heavily cratered surface with little sign of recent geological change.

Typical of minor planetary satellites, Proteus has an irregular shape. The "roundness" of a moon is determined by its mass. A massive body has enough gravity to compress it into a spherical form. Lacking sufficient gravity, smaller bodies wind up as what might best be described as "cosmic potatoes." Proteus appears to be on the boundary between the two situations; it's about as large as a moon can get while still lacking the gravity to pull it into a spherical form. It's one potato that didn't get gravitationally "mashed"!

TRITON

Pronounced: "TRY-tun"
Diameter: 2,704 kilometers (1,680 miles)
Mean Distance from Neptune: 354,760 kilometers (220,448 miles)
Orbital Period: 5.88 days (retrograde)
Period of Rotation: 5.88 days (synchronous)
Average Density: 2.1 g/cm^3
Surface Gravity, Compared to Earth: 8.0%

It wasn't a professional astronomer who discovered Neptune's largest moon, Triton. William Lassell, who spotted Triton on October 10, 1846, just 17 days after Neptune was discovered, was a brewmaster by trade. The fortune he made in the brewery business allowed him the luxury of pursuing his astronomical interests with a 61-centimeter (24-inch) reflecting telescope—the largest such instrument in his native England. Perhaps in lieu of celebrating his discovery with a glass of champagne, Lassell toasted the event with a mug of his finest beer!

Named after the son of Neptune, the God of the Sea, Triton is by far Neptune's largest satellite. It's three-fourths the size of our Moon and nearly seven times greater in diameter than Neptune's second-largest moon, Proteus. It's possible that Triton may contain as much as 99 percent of all the matter circling Neptune. Denser than many of the moons found in the outer solar system, Triton likely contains a higher percentage of rocky material.

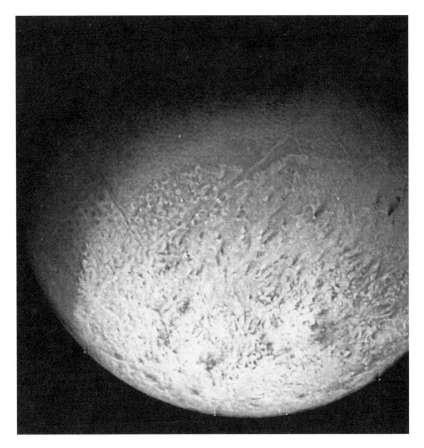

Figure 9.3 Triton. AP Photo/NASA.

Triton is unique among the solar system's large moons in that it orbits its home planet in a retrograde path opposite the direction of the planet's rotation. Not only does Triton orbit the "wrong" way, but the axis of that orbit is highly inclined to Neptune's equator. Combined with Neptune's 28° axis tilt, it places Triton's axis nearly 180° to the Sun. Like Uranus, Triton alternately turns each of its poles directly toward the Sun.

Its retrograde, highly inclined orbit, as well as similarities to what astronomers believe is its Pluto's composition, has led astronomers to postulate that Triton may have formed elsewhere in the solar system and strayed too close to Neptune. During its capture, Triton could have disrupted the orbits of existing satellites of Neptune. Such a scenario would explain the highly eccentric orbit of Nereid.

Triton's retrograde path around Neptune may lead to its ultimate demise. The orbit of this moon is slowly decaying. Astronomers estimate that billions of years from now, Triton will enter Neptune's Roche Zone, where it will break apart. The icy residue could form a bright ring, rivaling Saturn's.

When *Voyager 2* began imaging Neptune's moons in 1989, astronomers were already aware that planetary satellites were anything but dead, airless worlds. Jupiter's Io was alive with active volcanoes. Titan was veiled by a dense, methane-laced atmosphere. Miranda was a jumble of geologically fascinating terrain. Triton would prove to be a combination of the three.

Even before *Voyager 2* arrived at Neptune, astronomers had analyzed Triton with the spectroscope and found evidence for a thin nitrogen-methane atmosphere. *Voyager 2* confirmed those suspicions. It also showed that Triton has a highly reflective surface, comprised of rock-hard water ice topped by a coating of nitrogen and methane frost. Much of what little heat and light Titan receives from the Sun is reflected back into space. As a result, Triton has the coldest measured surface temperature (−235°C, or −391°F) of any body in the region of the solar system inhabited by the planets.

And what a strange-looking surface it is! Sparsely cratered, it shows signs of past geologic activity. Ridges and valleys indicate a period of melting and refreezing. Most remarkable is the portion of Triton's surface that is reminiscent of the rippled exterior of a cantaloupe. The agents that produced the "cantaloupe" terrain are still unknown.

Most surprising to the *Voyager 2* scientists is what appear to be geysers erupting on Triton's surface. Several dark plumes showed up in *Voyager 2* images, some rising 2 to 8 kilometers (1 to 5 miles) above the surface before being carried dozens of kilometers downwind by Triton's thin atmosphere. The geysers appear to be comprised of a combination of nitrogen and methane gases mixed in with dusty material.

How can geysers, which are associated with a heat source, exist on such a cold world? The answer may be found in seasonal patterns on Triton. Concentrated on Triton's south pole, which was facing sunward at the time of *Voyager 2*'s rendezvous, the geysers may be created as the ground heats up just enough so that subsurface nitrogen and methane liquefies. Under pressure, these gases erupt through fissures in Triton's crust to form the geysers. Since a season on Neptune (and Triton) lasts over 40 years, several generations of astronomers will have to monitor Triton to see if geysers begin erupting in the moon's northern hemisphere as winter there turns to summer.

NEREID

Pronounced: "NEER-ee-ed"
Diameter: 340 kilometers (210 miles)
Mean Distance from Neptune: 5,513,400 kilometers (3,426,025 miles)
highly eccentric: varies between 1,353,600 kilometers (841,100 miles) and 9,623,700 kilometers (5,980,200 miles)
Orbital Period: 360.14 days
Period of Rotation: 0.48 days?
Average Density: 1.5 g/cm^3 (estimated)
Surface Gravity, Compared to Earth: 0.7%

Nereid was the second of Neptune's satellites to be discovered. The honor goes to the Dutch-born astronomer Gerard Kuiper, who captured it on a photographic plate in 1949. What distinguishes Nereid from all the other moons in the solar system is its **orbital eccentricity**. In an ideal solar system, the orbit of each moon around its parent planet (and that planet's orbit around the Sun) would be a perfect circle. Instead, virtually all bodies in the solar system have orbits that are slightly elliptical. Nereid takes "elliptical" to an extreme. During its one-year orbit of Neptune, Nereid's distance from Neptune varies by a factor of more than seven.

Little is known about Nereid other than its orbit and size. Even *Voyager 2* added relatively little to our meager knowledge. When the craft passed through the Neptunian system, it never got closer to Nereid than 4,700,000 kilometers (2,920,580 miles).

Gerard Kuiper (1905–1973)

Dutch-born Gerard Kuiper did more than just discover Neptune's moon Nereid. He is considered to be the father of modern planetary astronomy. After studying astronomy at Leiden University, Kuiper moved to the United States. Settling in California in 1933, he worked at the Lick Observatory for two years before continuing on to the Harvard College Observatory where he met his wife.

In 1937, he took up a position at the Yerkes Observatory and University of Chicago, working there until 1960. While there, Kuiper discovered Nereid and Uranus's moon Miranda, and announced the presence of a thick methane atmosphere around Saturn's largest moon, Titan.

In 1960, Kuiper moved one last time to Tucson, Arizona, where he established the Lunar and Planetary Laboratory at the University of Arizona. He served as its director until his death in 1973. Kuiper realized that visible light provides only part of a planet's picture, but that much of the infrared light released by a planet is absorbed by our atmosphere before it ever reaches ground-based instruments. He was one of the first to propose high-altitude airborne studies of the planets in infrared light. The Kuiper Airborne Observatory (KAO), which detected the rings of Uranus during a stellar occultation, was named in Gerard Kuiper's honor.

NESO, MOST REMOTE MOON IN THE SOLAR SYSTEM

Pronounced: "NEE-soh"
Diameter: 60 kilometers (37 miles)
Mean Distance from Neptune: 49,285,000 kilometers (30,625,700 miles)
Orbital Period: 26.7 years
Period of Rotation: unknown
Average Density: 1.5 g/cm^3
Surface Gravity, Compared to Earth: unknown

Between 2002 and 2003, astronomers using CCD cameras with large ground-based telescopes discovered five distant satellites of Neptune. All are relatively small, with diameters between 30 and 60 kilometers (18.5 to

37 miles). These five moons are most likely captured objects, circling Neptune at vast distances, with orbital periods measured in years.

The remotest of the group is Neso, discovered by teams led by Matt Holman and Brett Gladman in 2002. This 60-kilometer (37-mile)-wide moon is farther from its home planet than any other moon in the solar system. Its distance from Neptune is over 125 times greater than the gap that separates the Earth from the Moon. In the time it takes Neso to complete a single orbit around Neptune, our Moon orbits Earth 325 times!

Will Pluto Crash into Neptune?

The orbit of Pluto is highly eccentric. At times, the dwarf planet drifts inside Neptune's orbit, where it remains for several decades before drifting back into the outer reaches of the solar system. Could the two ever collide?

Collisions of bodies in the solar system were commonplace billions of years ago when numerous planetesimals were accreting out of the solar nebula. Impacts still occur today, but they involve relatively small bodies like comets or meteoroids. Analysis of the orbits of Neptune and Pluto assure astronomers that while the two bodies may flirt, there's no chance for a cosmic marriage. Their orbits cross, but they don't intersect at the same spot.

Consider a crossroad between two well-traveled roads. Because they intersect, there is always the possibility of two automobiles colliding at that intersection if they arrive there at the same time (and if the drivers ignore the STOP signs). There are no STOP signs in the orbits of the planets. However, Pluto's orbit passes *above* Neptune's. Instead of a crossroads, we have an overpass that acts like a bridge carrying Pluto past Neptune, but safely out of harm's way.

NEPTUNE'S MOONS: THE MOST POPULAR FUTURE TOURIST SITES

So far in this book, we've highlighted future tourist sights on the moons of Jupiter, Saturn, and Uranus. We conclude our imaginary travelogue with a look at a vacation offering only the hardiest travelers would book.

1. Triton. So your friends are bragging about their trip to Antarctica, where the thermometer dipped to −100°F? Now you can one-up them with a vacation tour to the solar system's true deep freeze, Neptune's moon Triton. On Triton, temperatures hover around −390°F, colder than liquid nitrogen. In fact, the frost you see covering the ground on Triton is *frozen* nitrogen. The coldest day in Antarctica would be a heat wave on this world! While on Triton, you'll relax in luxurious heated living quarters, stationed near one of Triton's magnificent geysers. And don't forget a visit to the Triton Observatory, where a remote-controlled telescope will show you the planets from Neptune's unique vantage point as the solar system's most remote planet. You'll never forget your view of Earth and the Moon almost 3 billion miles away!

SEE FOR YOURSELF

Neptune, like Uranus, can be a hard-to-find target for the novice sky gazer. While it's true that the solar system's most remote planet can be glimpsed with binoculars, Neptune is still 10 times fainter than Uranus and looks like an ordinary 8th magnitude star. Unless you know exactly where to look, you'll likely miss it. Making a positive observation of Neptune requires the ability to find your way among the constellations using a finder chart.

The good news is that you won't have to make mathematical calculations like John Adams and Urbain Le Verrier did to find Neptune's location. Several annual astronomy guides, as well as the January issues of some of the popular astronomy magazines (see the Resource list in Appendix E) include finder charts for Uranus and Neptune.

Since binoculars won't magnify Neptune very much, you'll need a telescope and a power of at least 100X to see the planet's tiny disk. Imagine trying to look at a golf ball placed 3.7 kilometers (2.3 miles) away. That's how small Neptune's disk appears. No matter how hard you try, you won't see much more than a minute bluish dot.

If you're an astronomical neophyte who wants to see Neptune but fears getting lost in space, try contacting a local astronomy club to see if one of its members will show you the way. As was mentioned in the last chapter, you can log on to www.astronomyclubs.com to find a club near you. Neptune may not be the most spectacular telescopic sight, but it's nice to gaze at a planet that even the great astronomer Galileo Galilei missed.

WEB SITES

http://www.astronomycast.com/astronomy/episode-63-neptune.
  The script from an "Astronomycast" podcast highlighting Neptune. Includes links to the actual podcast, plus Neptune-related Web sites.
http://sse.jpl.nasa.gov/planets/profile.cfm?Object=Neptune.
  NASA Web site containing plenty of facts about Neptune.
http://www.nineplanets.org/neptune.html.
  A look at Neptune from a Nine (8) Planets Web site perspective.

10

Exoplanets: Jovian Planets Beyond Our Solar System

"HOT JUPITERS"

The discovery stunned the scientific world and turned a long-standing theory of planet formation on its head. On October 5, 1995, just two months before the much-anticipated arrival of the *Galileo Orbiter* at Jupiter, astronomers Michael Mayor and Didier Queloz of the University of Geneva announced the discovery of the first **extrasolar planet** in orbit around a Sun-like **main-sequence star**. The planet was too faint to see visually or even capture with a large CCD-equipped telescope. Its gravitational tug on the parent star, 51 Pegasi, betrayed its presence.

When astronomers determined the planet's orbit and size, they were in for a surprise. The planet, officially dedicated "51 Pegasi b," revolves around its parent star once every four days at a distance much closer than what separates Mercury from our Sun. More astounding than the planet's close proximity to its parent star is its size. Instead of being a tiny, low-mass terrestrial planet like those that orbit close to our Sun, 51 Pegasi b possesses the bulk of at least 150 Earths! Mayor and Queloz had discovered what would eventually be called a "**hot Jupiter**."

Planetary scientists were baffled. A body with gas giant credentials has no business being so close to its parent star. At the time 51 Pegasi was birthing planets, that region shouldn't have had enough material to form so massive a planet. The discovery in subsequent years of even more hot Jupiters forced astronomers to reevaluate the mechanics of planetary formation. The

Nebular Hypothesis, which seemed to fit the creation of the planets in our solar system like a glove, was now being put under the microscope. What happened here 4.6 billion years ago obviously wasn't the scenario for 51 Pegasi or for the other stars that serve as homes for hot Jupiters.

Why do so many hot Jupiters inhabit a forbidden zone for gas giant planets? Current wisdom says that these planets originally formed further out in the material-rich region of their home star's protoplanetary disk and were dragged inward by friction with the surrounding gases or perhaps by a gravitational encounter with another developing gas giant. Eventually, the planet settled into a new orbit close to the star, becoming a hot Jupiter. Because of its enormous mass, it possessed enough gravity to counteract the attempts by the star's radiant energy to strip it of its gaseous envelope.

The story of exoplanets and hot Jupiters takes us back to the beginning of 1992. At that time, there were eight known planets in the entire universe, namely, the ones that orbit our Sun. On a few earlier occasions, astronomers had claimed to have detected planets in orbit around some of the nearest stars. As fast as astronomers announced these findings, follow-up observations showed the data were incorrect or misinterpreted. While it might seem arrogant and foolish to assume that our Sun is the only star in the universe to be the center of a planetary system, none had yet been detected.

Why did astronomers have trouble finding planets beyond our solar system? We can understand the difficulty by recalling the situation surrounding the discovery of the moons circling the outer planets. Because a planet is much larger than its moons, it appears brighter in the sky. The overpowering glare reduces the visibility of its moons. The situation with extrasolar planets is magnified for two reasons. First of all, extrasolar planets are a thousand times more remote than the outer planets and their moons. Second, a typical star outshines its planets a million fold. Direct visual detection of an exoplanet is like spotting a mosquito flying around a searchlight dozens of kilometers away!

By the early 1990s, astronomers had at last developed technologies with the sensitivity to capture those searchlight-circling mosquitoes. In 1992, a group of astronomers presented conclusive evidence that planets had been detected around the pulsar PSR 1257+12. A pulsar is the superdense core of a dead star—all that remains when a massive star undergoes a violent supernova explosion. This stellar corpse rotates rapidly, releasing radio waves at precise intervals. Irregularities in the radio emissions of PSR 1257+12 betrayed the presence of several planetary bodies in orbit around the dead star. As they circled the pulsar, these planets gave it gravitational tugs that caused periodic fluctuations in the radio signals it was emitting. The planets discovered around pulsar PSR 1257+12 had either formed from the material ejected by the supernova or they were the remains of planets that had already been in existence at the time of the explosion.

The technique used by Mayor and Queloz to detect 51 Pegasi b is what astronomers call the **radial-velocity method**. It's the means by which most extrasolar planets have since been discovered. As a planet orbits a star, its gravitational tug causes the star to wobble slightly. The wobble causes a periodic Doppler shift in the star's spectral lines.

Another common approach for tracking down extrasolar planets is the **transit method**. If the planet's orbit is edge-on to our line of sight, it will periodically pass in front of (transit) the parent star. Sensitive detection equipment will record a slight dip in the star's overall brightness. Careful analysis of the event will yield information about the planet's size and orbit.

One of the more exotic methods of discovering extrasolar planets is **microlensing**. According to Einstein's General Theory of Relativity, a strong gravity field will bend a passing beam of light. If a star, during its wanderings around our galaxy, were to pass in front of a more distant star, it would have a microlensing effect on the light from that star. Any planets orbiting the closer star would produce additional microlensing that could be detected. In this way, astronomers have discovered a handful of exoplanets. A problem with microlensing is that it's a "one shot" deal. Once the star continues past the background star, the event and the opportunity to confirm the data or discover more planets is over.

In November 2008, two separate groups of astronomers announced the first confirmed direct images of extrasolar planets. A team led by Dr. Christian Marois of the Herzborg Institute of Astrophysics used a pair of huge reflecting telescopes on Mauna Kea to image three planets in orbit around the star HR 8799. All three are more massive than Jupiter and are classical gas giants with orbits far from their parent star. Meanwhile, a planet with three times Jupiter's mass was found orbiting the bright star Fomalhaut. Dr. Paul Kalas of the University of California at Berkeley used the Hubble Space Telescope and a special filter to block Fomalhaut's glare. The planet, Fomalhaut b, is another textbook gas giant, orbiting its parent star at a distance of about one hundred astronomical units.

Does the preponderance of super massive planets thus far discovered mean that terrestrial-sized planets are nonexistent outside our solar system? Hardly! Our current means of detection lack the necessary sensitivity to uncover Earth-sized planets.

In recent years, as the sensitivity of detection devices continues to improve, astronomers have begun to ferret out so-called "super earths"—planets with masses several times Earth's. It is quite likely that by the time this book goes to press, astronomers will have announced the discovery of an Earth-sized planet. Early in 2009, NASA launched a specially designed telescope that will monitor the brightness of 100,000 stars in an effort to detect planetary transits. Called the Kepler mission, it will be able to capture Earth-sized planets as well as Jovian-sized planets. In the years to come, science will have a more complete picture of the families of planets orbiting the nearby stars, allowing for a better understanding of the role Jovian planets, both classical gas giants and hot Jupiters, play in planetary systems.

WEB SITES

http://planetquest.jpl.nasa.gov.

  NASA's information-packed Web site on extrasolar planet research.

http://exoplanet.eu.

  This Web site includes an up-to-date listing, along with data, of all currently known extrasolar planets.

http://kepler.nasa.gov/about.

  This NASA Web site details the Kepler mission to search for extrasolar planets.

11

Voyager 2's Grand Tour

January 1, 2009: *Voyager 2* is traveling at a speed of 15.5 kilometers (9.3 miles) per second away from the Sun. At this speed, the craft could circle the Earth in 45 minutes. It is now 88 Astronomical Units from Earth, nearly three times the distance of Neptune. The aging craft continues to send back information about the remote reaches of the solar system. Radio waves from *Voyager 2* take about 12.5 hours to reach Earth.

Back on Earth, Michelle is an astronomer who teaches space science at a local college. Now in her mid-30s, she often takes her children outside to look at the stars.

THE GRAND TOUR

Voyager 2 wasn't the first spacecraft to explore the outer planets. *Pioneer 10* flew past Jupiter in 1973, followed by *Pioneer 11*, which reached Jupiter in 1974 and Saturn in 1979. *Voyager 1*, launched weeks after *Voyager 2* but set on a faster course, arrived at Jupiter and Saturn ahead of its twin. What makes *Voyager 2* unique is that it is the only craft to date to complete a Grand Tour of all four outer planets.

You might wonder why scientists would want to launch robotic space probes to Jupiter, Saturn, Uranus, and Neptune when we have an arsenal of telescopes here on the Earth. The answer is obvious to anyone who has ever looked through a telescope. Under the highest possible magnifications, the planets present tantalizingly small disks. Close-up detail of their moons is all but impossible, even with the Hubble Space Telescope or a large

159

ground-based telescope enhanced with adaptive optics. Besides providing a close-up view of a distant world, a space probe can better analyze its chemical composition and map its magnetosphere.

Despite the obvious advantages, a space probe has one major disadvantage. Cost. The design of a probe and the rocket that will send it on its way requires thousands of man-hours. Construction means expensive materials and adds even more man-hours. During the mission, the man-hours continue to pile up as people continually monitor and control the probe during its voyage, then collect and analyze the data once the probe arrives at its destination. The price tag for a typical interplanetary mission may run in the billions of dollars.

Scientists understand the expense of flyby missions and do their best to keep costs low. One way is to explore several planets with the same craft. There is a hidden bonus to such a mission. As the craft approaches the first planet, that planet's gravity begins to pull on the craft, causing it to speed up. Mission controllers on the ground carefully adjust the craft's trajectory so that it will race past the planet and then "slingshot" onward to the next planet. The gravity boost cuts costs two ways by saving fuel and reducing the time of the mission.

The 1970s afforded a literal once-in-a-lifetime opportunity to send a space probe to all four of the outer planets. Once every 175 years, Jupiter, Saturn, Uranus, and Neptune are roughly aligned, making a Grand Tour of the quartet possible. While a direct flight to Neptune might take four decades, a probe getting gravity assists from Jupiter, Saturn, and Uranus could make the trip in 12 years. Space scientists were excited about the possibility, but there was one problem. For economic reasons, Congress rejected the proposed mission in 1971. The Grand Tour got a second life a year later, when it was decided to cut costs by modifying an existing *Mariner* probe to make the trek. Part of the reason for resurrecting the Grand Tour may have been the realization that the last time the four outer planets were aligned was during the presidency of Thomas Jefferson! Eventually, two *Mariner* probes, renamed *Voyager 1* and *2* were targeted for use.

Voyager 2 would be launched first, with *Voyager 1* sent a few weeks later on a faster, more direct route. The primary mission of *Voyager 1* was to collect data from flybys of Jupiter and Saturn. *Voyager 2* was designed to observe Jupiter and Saturn as well, but then continue on to Uranus and Neptune. This extended mission would happen *only* if *Voyager 1* had a successful Saturn encounter. If *Voyager 1* failed at Saturn, *Voyager 2*'s path would be altered to get a better view of the Ringed Planet and its moons. But, the altered path would not allow *Voyager 2* to continue on to Uranus and Neptune. Fortunately, *Voyager 1* obtained spectacular results during its 1980 flyby of Saturn. The door was open for *Voyager 2* and its Grand Tour.

BEATING THE ODDS

Imagine the difficulty in planning *Voyager 2*'s Grand Tour mission to the outer planets. You need to modify a *Mariner* probe so it can safely negotiate a voyage of billions of kilometers through the hazardous environment of space. Among other things, it needs to survive extremes in temperature, a journey through the asteroid belt, and zones of high radiation surrounding the planets it visits. It must be able to operate continually during a trek lasting at least a dozen years. On-board computers and scientific instruments, from spectrometers to cameras, must likewise be durable. Above all, you must ensure a constant and reliable two-way communication between yourself and the probe.

Next you have to carefully plan *Voyager*'s trajectory. At the time of launch, you direct the craft not where Jupiter is, but where Jupiter will be two years later when the craft arrives. You must consider the positions of Jupiter's moons at the time of the encounter. If they lie on the opposite side of the planet when *Voyager 2* arrives, you won't get the desired close-up images. You also need to program *Voyager 2* to enter Jupiter's gravity field at just the right position so when it sweeps past Jupiter, it will get a gravitational boost toward Saturn—not where Saturn is, but where Saturn will be at the time *Voyager 2* arrives several years later. Again, you want to be sure that Saturn's moons of interest are well-placed for a photo op, and that the craft hits Saturn's field of gravity just right for a send-off to Uranus. The scenario repeats at Uranus and Neptune. Knowing the complexity of

Figure 11.1 *Voyager 2*. NASA.

Voyager 2's trajectory, we can appreciate the true meaning of the term *rocket scientist* for describing someone of high brainpower!

Obviously it wasn't a single rocket scientist who handled all the details of the Grand Tour of *Voyager 2*. Teams of skilled and dedicated scientists, technicians, and laborers saw the program through from beginning to end. Still, problems arose that threatened this one-shot mission.

About seven months after launch, a power surge destroyed *Voyager 2*'s primary radio receiver and damaged the backup to the point that it could only receive signals in a limited range of radio wavelengths. Fortunately, NASA scientists were able to adjust the frequency of the command signals so *Voyager 2* could "hear" them.

All went well at Jupiter, but troubles arose at Saturn. A scan platform that allowed the *Voyager 2* cameras to pan an object as the craft sped by jammed. To understand the severity of this problem, imagine trying to take a picture of a roadside sign from the open window of a speeding car. If you just hold the camera steady and snap the shutter as the object crosses the view finder, you'll probably get a blurry picture. But if you aim the camera at the sign and turn the camera to keep it centered in the view finder as you snap the shutter, you'll get a sharper picture. At Saturn, *Voyager* had already taken a large number of pictures when the scan platform locked. The prospects for Uranus looked bleak. Again, the ingenuity of the *Voyager* scientists saved the day. They programmed the entire craft to spin slowly to keep a target centered on the cameras. The *Voyager 2* Uranus encounter was saved.

One of the fascinating aspects of the *Voyager 2* mission was the way teams of scientists on Earth were able to reprogram the probe to solve problems and even improve its capabilities as it sped past the planets. By the time *Voyager 2* reached Neptune, and despite its ailments, the probe was in some ways a better machine than the one that blasted off from Cape Canaveral a dozen years earlier. Photographs of Neptune, its rings, and moons were astounding. Our Grand Tour emissary from Earth had defied the odds and survived a voyage to the four outermost planets in the solar system. Mission accomplished!

WEB SITES

http://voyager.jpl.nasa.gov.
> NASA's *Voyager* Web site. Contains a wealth of information about the *Voyager* missions.

http://heavens-above.com/solar-escape.asp?
> A Web site that provides up-to-date information on the current status of *Voyager 2* and other space probes destined to leave the solar system.

12

The Outer Planets: A Glance
Back and a Look Ahead

January 1, 2025: It's a somber New Year's Day at the Jet Propulsion Lab in Pasadena. *Voyager 2* has been silent for the past few weeks. The amazing interplanetary probe that brought the Earth its first close-up photos of Uranus and Neptune is officially dead. The craft outlived many of the scientists who created it back in the early 1970s.

Michelle steps outside to take a quick look at the night sky. Her daughter, now a college astronomy major, is with her, but Michelle feels strangely alone. An old friend has died. "Rest in peace," she whispers before returning inside.

A LOOK BACK

From its launch at Cape Canaveral to its flyby of Neptune was a journey of a dozen years and 4.5 billion kilometers (2.8 billion miles) for *Voyager 2*. For the humans who created *Voyager 2*, it was merely the latest stage of a journey of understanding that had begun thousands of years earlier.

Humans have come a long way from the prehistoric skygazers who first recognized Jupiter and Saturn as slow-moving "wandering stars." In just a few thousand years—a cosmic blink of the eye, when you consider the age of the solar system—we have evolved from primitive beings who regarded the planets with superstition to scientific intellects who study them with genuine curiosity.

Long before humans placed glass lenses in a tube to obtain close-up views of the Moon and beyond, early astronomers had plotted the motions of the Sun, Moon, and planets to create calendars and produce early models of the solar system. The telescope helped us refine our ideas about the organization and makeup of the solar system. We learned that the outer planets are quite different from our Earth. When the telescope proved inadequate to ferret out the mysteries of the outer planets, human ingenuity provided new tools like the spectroscope and camera to do the job. In recent decades, the radio telescope, satellites, space probes, and an orbiting space telescope have added volumes to our store of knowledge about Jupiter, Saturn, Uranus, and Neptune.

Voyager 2 and the other probes that have explored the outer planets answered a lot of questions about these remote worlds and their moons. As with all aspects of science, each question answered raises many more. *Voyager* taught us that the surface of Europa appears to be composed of frozen water ice. The question now is whether or not a sea of liquid water lies beneath this icy crust. If so, could life exist there? We have more knowledge about the outer planets than Galileo, but we also have more questions.

THE PRESENT

The *Cassini Orbiter,* which arrived at Saturn in 2004 to begin a four-year mission, made amazing and unexpected discoveries about the Ringed Planet and its moons. In 2008, its tour of duty was extended for another two years. As this book goes to press, *Cassini* may be imaging a storm in Saturn's atmosphere, studying the geysers of Enceladus, or mapping the terrain of Titan's surface. We can be sure that more amazing discoveries will be sent back to Earth.

Back on Earth, huge ground-based telescopes, whose capabilities are enhanced by adaptive optics and CCD technology, continue to keep an eye on the outer planets, especially Uranus and Neptune. With no probes planned for these planets in the foreseeable future, this is the only way astronomers can monitor their changing weather patterns. Orbiting above the Earth's often turbulent atmosphere, the Hubble Space Telescope is occasionally called upon to augment the ground-based observations.

A LOOK AHEAD

Just a half century ago, the astronomer's tool bag consisted mainly of the telescope, spectroscope, and camera. The Space Age was in its infancy, as was radio astronomy. The tools astronomers will have 50 years from now cannot even be imagined. New designs in rocket propulsion may send probes to the outer planets that arrive in a matter of months, not years. We won't have to worry about the limitations our atmosphere puts on ground-based optical telescopes. Huge automated observatories on the Moon will provide images of the outer planets that rival what our present-day space probes have returned. Indeed, we may be receiving detailed images not only of Jupiter, but of the hot Jupiters that orbit nearby stars.

Closer to the present, several missions are planned that will enhance our knowledge of the outer planets. The Hubble Space Telescope is due for repair work and upgrading early in 2009. Some time around 2013, the James Webb Space Telescope will be sent into space. Designed to study infrared energy, this 6.5-meter (21.3-foot) telescope will provide critically needed observations of the outer planets.

Jupiter is due for a serious physical examination when NASA's Juno Mission, scheduled for launch in 2011, arrives seven years later. Juno will orbit Jupiter for one year, carefully monitoring its gravity and magnetic fields and analyzing deep portions of its atmosphere, all in an attempt to glean information about the giant planet's internal structure. Knowing what lies beneath Jupiter's cloud-laden atmosphere should give astronomers clues about the planet's formation and the birth of the solar system in general.

Does life exist elsewhere in the universe? That's one of the most compelling questions facing astronomers. The answer might be right on our doorstep, cosmically speaking. The *Europa Explorer* mission, still in the planning stages, would perform a reconnaissance of Jupiter's ocean moon Europa. Plans call for a 2015 launch, with arrival at Jupiter half a dozen years later. After orbiting Jupiter and its moons *Cassini*-style for two years, the *Europa Explorer* would fall into a tight one-year orbit around Europa, studying the moon at extremely close range. The mission could lead to a soft landing by a proposed Europa Astrobiology Lander. The device would actually drill through Europa's icy crust and launch a submersible that would "taste" the chemical composition of Europa's subterranean ocean and look for possible life forms.

Beyond that, it's only a matter of time before human explorers go to Jupiter and conduct firsthand explorations. Voyages to those colossal worlds that exist far from the Sun will continue for as long as humans have questions that need answering. Enjoy the ride!

WEB SITES

http://hubblesite.org and http://hubble.nasa.gov.
> A pair of NASA-based Web sites that offer information about the Hubble Space Telescope.

http://www.jwst.nasa.gov.
> NASA's James Webb Space Telescope Web site. If you want to know what's happening with this telescope, check this site.

http://www.keckobservatory.org.
> A Web site that describes the work being done at one of the world's largest ground-based telescopes.

Appendix A: Planetary Data Tables

Planet	Distance from Sun (au)	Equatorial Diameter (km)	Mass (Earth = 1)	Density (g/cm³)	Period of Rotation	Period of Revolution	Moons	Rings
The Terrestrial Planets								
Mercury	0.4	4,879	0.06	5.4	58.6 days	88.0 days	0	N
Venus	0.7	12,104	0.82	5.2	243 days*	224.7 days	0	N
Earth	1.0	12,756	1	5.5	23h 56m	365.2 days	1	N
Mars	1.5	6,792	0.11	3.9	24h 37m	686.9 days	2	N
The Jovian Planets								
Jupiter	5.2	142,984	317.8	1.3	9h 56m	11.9 years	63	Y
Saturn	9.5	120,536	95.2	0.7	10h 39m	29.5 years	60	Y
Uranus	19.2	51,100	14.5	1.3	17h 14m	84.0 years	27	Y
Neptune	30.1	49,538	17.1	1.6	16h 7m	164.8 years	13	Y

Note: * = retrograde

Appendix B: The Chemistry of the Outer Planets: Elements and Compounds

Outer space may be alien, but the materials that make up the bodies in the universe aren't. The elements we find here on Earth exist throughout space—in stars, planets and their moons, and comets. The only difference from their earthly forms might be the states in which they exist. The following list describes the nature of the "stuff" of the outer planets, their rings, and moons.

Ammonia, NH_3: A colorless, pungent gas. Among its sources on Earth is the decay of animal and planet matter. Often found in household cleaners and fertilizers. Ammonia is present in the atmospheres of the gas giant planets.

Ammonia hydrosulfide, $(NH_4)SH$: A pungent gas that is found in so-called "stink bombs." Occurs in the atmospheres of the Jovian planets.

Carbon dioxide (CO_2): A gas found in Earth's atmosphere, commonly used by plants during photosynthesis. Also found in the atmospheres of some of the planets and frozen as "dry ice" on the surfaces of a few of the moons of the Jovian planets.

Electron: A negatively charged particle that occurs in the outer shells of atoms. Free-flowing electrons generate electric current, which, in turn, creates magnetic fields. Electrons are found in the solar wind that emanates from the Sun.

Ethane, C_2H_6: A colorless, odorless hydrocarbon that exists as a gas under normal temperatures and pressures. Ethane is found in natural gas. It also makes up a small portion of the atmospheres of the Jovian planets, and possibly exists in liquid form in lakes on Titan's surface.

Gas: Describes a state of matter that has neither a definite shape nor volume. Also refers to the light substances, mainly hydrogen and helium, that make up the gas giants Jupiter and Saturn. These can exist in a gaseous state or as a liquid or solid under the extremes of temperature and pressure found in the depths of the Jovian planets or on the surfaces of their moons.

Helium, He: The second most common element in the universe, commonly used on Earth for balloons and dirigibles. A major component of the Sun and Jovian planets.

Hydrogen (molecular), H_2: The most common and lightest element in the universe. Hydrogen atoms generally occur in pairs (molecular hydrogen). Hydrogen is a primary component of the Sun and the Jovian planets. In the Jovian planets, it behaves as a liquid metal in their high-pressure interiors.

Hydrogen sulfide, H_2S: A colorless gas with a distinct "rotten egg" smell, produced on Earth when bacteria break down sulfates in the absence of oxygen, as in swamps or sewers. Also a byproduct of volcanoes and hot springs. A minor component of the atmospheres of the Jovian planets.

Ice: A common description for the solid condition of water under freezing conditions on Earth. Also describes the substances (mainly water, ammonia, and methane) that exist as ices in the cold temperatures around the Jovian planets. A primary ingredient in the moons of the outer planets.

Ion: An atomic particle that carries an electric charge, either by losing or gaining one or more electrons. Commonly found in the magnetospheres of the planets.

Iron, Fe: A metal commonly occurring in the Earth's crust, originally formed in the cores of red supergiant stars that went supernova. Also found in a solid or liquid state in the cores of most of the planets.

Metal: A relatively dense substance found commonly on the surfaces and in the cores of the terrestrial planets and in lesser quantities in the cores of the Jovian planets.

Methane, CH_4: A colorless, odorless gas that is the prime ingredient of natural gas. On Earth, methane forms naturally by the bacterial decomposition of plant and animal matter that, at the bottom of lakes and ponds, forms "swamp gas." Methane is present in the atmospheres of all the Jovian planets (especially Uranus and Neptune), as well as the atmosphere of Saturn's moon, Titan.

Nickel, Ni: With iron, a metal commonly found in the crusts and cores of the dense terrestrial planets and, to a lesser extent, in the cores of the Jovian planets.

Nitrogen (molecular), N_2: A colorless, odorless gas whose atoms, like hydrogen's, naturally occur in pairs. Nitrogen is the most common element in the atmospheres of Earth and Saturn's moon Titan.

Organic compounds: A material that contains carbon and hydrogen and usually other elements such as nitrogen, sulfur, and oxygen. Organic compounds can be found in nature, or they can be synthesized in the laboratory.

Oxygen (molecular), O_2: An odorless, tasteless gas found in abundance in Earth's atmosphere. Found in trace amounts around some of the moons of the Jovian planets as a result of the breakdown of water ice on their surfaces.

Plasma: The fourth and most common state of matter, the others being solid, liquid, and gas. Plasma consists of a collection of free-moving electrons and ions, often under conditions of high heat. Stars are basically made up of plasma, as are the materials in planetary magnetospheres.

Silicate. A generic term describing the rocky material comprising much of the crusts of the terrestrial planets. Also found in many of the planetary satellites and the cores of the Jovian planets.

Sodium, Na: A highly reactive element found in ordinary table salt (NaCl). Occurs around Jupiter's moon Io.

Sulfur, S: A yellowish element commonly found near volcanic deposits in the Earth's crust. Also found in abundance on the surface of Jupiter's volcanic moon Io.

Volatiles: Substances with low boiling points (like water or methane) that could only exist in solid form in the area inhabited by the Jovian planets.

Appendix C: Tools of Discovery: The Electromagnetic Spectrum

THE NAKED EYE

We might not think of the unaided eye as much of an astronomical tool, but it is arguably the most important of all. Much basic information about the cosmos was gleaned by the eyes of prehistoric humans long before the invention of the telescope. Most of the named stars and constellations, as well as the discovery of five "wandering stars," including Jupiter and Saturn, came by way of the unaided eye. Even today, as the eye has been replaced by far more sensitive tools, its value to the astronomer cannot be overstated.

THE ASTROLABE AND ARMILLARY SPHERE

From the time of the ancient Greeks, astronomers have relied on sighting devices to obtain accurate positions of celestial bodies, in particular, the planets. Among the earliest were the astrolabe and the armillary sphere. The use of these devices during Renaissance times allowed for accurate reckoning of the positions of the planets. These data permitted astronomers like Kepler to calculate the planets' orbits.

THE TELESCOPE

Although there are claims that Roger Bacon fashioned a telescope sometime in the thirteenth century, most historians agree that the first working telescope was assembled by the Dutch lens maker Hans Lippershay in 1608. Fashioned from two lenses—an outer, or objective, lens that captured the light from a distant object and assembled it into an image, and an inner lens, or eyepiece, that magnified the image—it was the forerunner of what we today call the refracting telescope.

A half century later, a telescope that collects light with a concave mirror was introduced. The so-called reflecting telescope, capable of being constructed in truly gargantuan sizes, became the mainstay of astronomy during much of the twentieth century.

Modern-day telescopes are true works of wonder. With computer-controlled mirrors that flex to cancel out the effects of atmospheric turbulence (adaptive optics), huge ground-based reflecting telescopes produce crisp images that rival those of space telescopes like the Hubble.

THE PHOTOGRAPHIC PLATE

The human eye is an amazing device, but it has its limits. Here's where the photographic plate comes in. Astrophotography had its genesis in the mid-1800s. Faster and more accurate than hand sketches and more sensitive to faint light than the human eye, the photographic plate quickly became a major tool of astronomy. In the latter part of the twentieth century, the photographic plate was replaced by the much faster and even more sensitive charge-coupled device (CCD).

THE SPECTROSCOPE

What is a star made of? The spectroscope helps astronomers find out. Like the photographic plate, the spectroscope made its appearance in the mid-1800s. Starlight passing through the spectroscope's prism or diffraction grating is separated into its component colors (a spectrum). Lines in that spectrum are produced by the various elements in the star. The spectroscope can also indicate motion toward or away from the observer by the Doppler shift of the spectral lines.

THE RADIO TELESCOPE

The radio telescope was an accidental discovery. In the 1930s, Bell Laboratory scientist Karl Jansky was looking for the source of static that was disrupting radio communication. Using a primitive radio receiver, Jansky detected radio energy from discreet sources in space. After World War II, astronomers improved Jansky's detection device to produce the world's first radio telescopes.

THE COMPUTER

It took the astronomer Johannes Kepler several years to calculate Mars's orbit, thereby proving that the planets circle the Sun. Calculating the position of Neptune was also a years-long task for John Couch Adams and Urbain Le Verrier. Had these mathematicians had a computer, the task might have been completed in a matter of hours! Since the mid-1900s, the computer has helped astronomers with tasks as varied as calculating the orbit of a comet to controlling a telescope or space probe.

SATELLITES AND SPACE PROBES

The arrival of the Space Age with the launch of the Soviet Union's Sputnik satellite in October 1957 ushered in a new era of astronomy. Satellites in Earth orbit could study forms of energy, like ultraviolet and infrared, normally blocked by our atmosphere. Space probes like *Voyager 1* and *2* allowed astronomers the luxury of close-up studies of bodies in the solar system.

Appendix D: The Outer Planets: Timeline of Discovery

Prehistory	Early humans discover five wandering "stars." Among them are Jupiter and Saturn.
ca. 2000 BC	Babylonians develop astrology; observe motions of Jupiter and Saturn.
140 AD	Ptolemy introduces a geocentric model of the solar system.
1543	Copernicus publishes heliocentric model of the solar system.
1608	First telescopes made in Holland.
1610	Galileo uses telescope to discover Jupiter's four major moons; observes, but fails to identify, Saturn's rings.
1612	Galileo unknowingly observes and sketches Neptune in the same field of view as Jupiter.
1655	Huygens discovers Saturn's largest moon, Titan, and correctly identifies the basic nature of Saturn's rings.
1665	Cassini reports the discovery of a large red spot on Jupiter, which remains visible for nearly 50 years.
1671	Cassini discovers Saturn's second moon, Iapetus.
1672	Cassini discovers Saturn's third moon, Rhea.
1675	Cassini discovers gap in Saturn's ring.
1684	Cassini discovers Saturn's fourth and fifth moons, Tethys and Dione.
1687	Newton publishes the *Principia*, in which he describes the force of gravity. It explains what holds the planets in their orbits.
1691	Cassini observes Jupiter's flattened disk.
1755	Kant proposes that stars and planets form from collapsing nebulae.
1781	Herschel discovers Uranus.
1787	Herschel discovers the first two moons of Uranus, Titania and Oberon.
1789	Herschel discovers Saturn's sixth and seventh moons, Enceladus and Mimas.

1796	Laplace publishes Nebular Hypothesis of solar system formation.
1830	Jupiter's Great Red Spot regains prominence; it has been observed continuously since.
1838	Earliest attempt at astrophotography, a picture of the Moon taken by Daguerre.
1843	Adams begins calculations of location of hypothetical planet beyond Uranus.
1845	Le Verrier begins calculations of location of hypothetical trans-Uranian planet.
1846	Galle discovers Neptune, using Le Verrier's calculations. Lassell discovers Neptune's largest moon, Triton.
1848	Lassell and Bond independently discover Saturn's eighth moon, Hyperion.
1850	W. C. Bond and his son discover Saturn's C (Crepe) Ring.
1851	Lassell discovers Uranus's third and fourth moons, Ariel and Umbriel.
1859	Maxwell provides mathematical proof that Saturn's rings are not solid, but are made up of a multitude of tiny particles.
1863	First astronomical use of a spectroscope. Huggins determines that the Sun is composed primarily of hydrogen.
1892	Barnard discovers Jupiter's fifth moon, Amalthea.
1899	Pickering discovers Saturn's ninth moon, Phoebe, on a photographic plate taken the year before. It is the first moon discovered with this method.
1904	Perrine discovers Jupiter's sixth moon, Himalia.
1905	Perrine discovers Jupiter's seventh moon, Elara.
1908	Melotte discovers Jupiter's eighth moon, Pasiphae.
1914	Nicholson discovers Jupiter's ninth moon, Sinope.
1933	Birth of radio astronomy, as Jansky reports the discovery of radio waves from outer space.
1935	Methane spectroscopically discovered in the atmospheres of Jupiter, Saturn, Uranus, and Neptune.
1938	Nicholson discovers Jupiter's tenth and eleventh moons, Lysithea and Carme.
1948	Kuiper discovers Uranus's fifth moon, Miranda.
1949	Kuiper discovers Neptune's second moon, Nereid.
1951	Nicholson discovers Jupiter's twelfth moon, Ananke.
1966	Dollfus discovers Saturn's tenth moon, Janus.
1969	Experiments at Bell Laboratories result in the beginnings of charge-coupled device (CCD) photography.
1972	*Pioneer 10* launched toward Jupiter.
1973	*Pioneer 11* launched toward Jupiter, will continue on to Saturn. *Pioneer 10* reaches Jupiter.
1974	*Pioneer 11* reaches Jupiter, continues towards Saturn.

1977	Rings of Uranus discovered during stellar occultation. *Voyager 2* launched toward Jupiter. *Voyager 2* will continue onward to Saturn, Uranus, and Neptune. *Voyager 1* launched toward Jupiter, will continue on to Saturn.
1979	*Voyager 1* and then *Voyager 2* arrive at Jupiter. *Pioneer 11* reaches Saturn.
1980	*Voyager 1* reaches Saturn, heads out of solar system.
1981	*Voyager 2* reaches Saturn, continues on to Uranus.
1984	Partial ring system of Neptune discovered during stellar occultation.
1986	*Voyager 2* arrives at Uranus. Space shuttle Challenger explodes after liftoff just four days later.
1989	*Voyager 2* arrives at Neptune. *Galileo* mission to Jupiter launched from the Space Shuttle Atlantis. The 3.6-meter (11.8-foot) telescope at the La Silla Observatory in Chile is the world's first to use adaptive optics.
1994	Impact of comet Shoemaker/Levy 9 with Jupiter observed by astronomers around the world.
1995	First "hot Jupiter," 51 Pegasi b, is discovered.
1995–2003	*Galileo* mission at Jupiter.
1997	*Cassini-Huygens* mission to Saturn launched.
1999–2007	Various groups of astronomers using CCD technology and adaptive optics on large, ground-based telescopes discover dozens of small moons around the four outer planets.
2004–2008	*Cassini* mission at Saturn. Due to its successes, the mission is extended two years until 2010.
2006	The International Astronomical Union (IAU) officially defines the characteristics of a planet; Pluto is demoted to "dwarf planet" status, making Neptune the remotest planet in the solar system.

Appendix E: Resources: Organizations, Publications, and Web Sites

ORGANIZATIONS

Association of Lunar and Planetary Observers (ALPO)
http://alpo-astronomy.org
Walter H. Hass, founder and director emeritus
2225 Thomas Drive
Las Cruces, NM 88001
E-mail: haasw@zianet.com
An organization devoted to advancing and conducting work by both professional and amateur astronomers who share an interest in solar system observations.

Planetary Society
http://www.planetary.org
65 North Catalina Avenue
Pasadena, CA 91106-2301
Phone: 626-793-5100
Fax: 626-793-5528
E-mail: tps@planetary.org
The world's largest space interest group, the Planetary Society is dedicated to the exploration of the solar system.

PUBLICATIONS

Astronomy
http://www.astronomy.com
Kalmbach Publishing Co.
21027 Crossroads Circle
P.O. Box 1612
Waukesha, WI 53187-1612
Phone: 800-533-6644 (U.S.A., Canada); +1 262-796-8776 ext. 421 (international)
E-mail: letters@astronomy.com
The world's largest-circulation astronomy magazine. Monthly issues contain astronomy-related articles and news, plus a sky calendar and chart showing the locations of the planets.

Discover
http://discovermagazine.com
Discover Magazine
90 Fifth Ave.
New York, NY 10011
Phone: 212-624-4774
E-mail: editorial@discovermagazine.com
A monthly science magazine that often contains astronomy-related articles.

Scientific American
http://www.sciam.com
Scientific American, Inc.
415 Madison Ave.
New York, NY 10017
Phone: 212-451-8200
E-mail: experts@sciam.com
Excellent science magazine that, like *Discover*, often contains articles dealing with astronomical topics.

Sky and Telescope
http://www.skyandtelescope.com
Sky Publishing
90 Sherman St.
Cambridge, MA 02140
Phone: 866-644-1377 (U.S.A., Canada); +1 617-864-7360 (international)
E-mail: info@skyandtelescope.com
This venerable magazine began in 1941. Similar format to *Astronomy*.

WEB SITES

Astronomy Cast
http://www.astronomycast.com.
A weekly podcast from astronomers Fraser Cain and Pamela Gay. Includes archives from past programs, as well as links to useful Web sites.

NASA
http://www.nasa.gov.
The primary Web site for the United States' space agency NASA. A prime source of data and information for this book. Of interest are the following NASA sites:
http://www.nasa.gov/mission_pages/cassini/main/index.html.
Web page for the *Cassini-Huygens* mission to Saturn
http://solarsystem.nasa.gov/planets/index.cfm.
Fact files for the planets and their satellites.

Nine (8) Planets
http://www.nineplanets.org.
Bill Arnett's informative Web site. Includes pertinent data on the planets and their moons.

Sky Maps
http://www.skymaps.com.
A free monthly sky calendar and map. A useful aid to finding Jupiter and Saturn in the night sky.

Scott Sheppard's Giant Planet Satellite and Moon Page
http://www.dtm.ciw.edu/sheppard/satellites.
Up-to-date data on the satellites that orbit the Jovian planets, compiled by the astronomer who has discovered the most planetary moons.

Glossary

Adaptive optics. A technique involving large reflecting telescopes with flexible, computer-controlled mirrors that compensate for atmospheric turbulence, producing a sharp image.

Albedo. A measure of the reflectivity of a nonluminous body like a planet or moon. Values range from 0 for a dark body to 1 for one that reflects 100 percent of the light striking its surface.

Arcsecond. An angular measure equal to 1/3,600 of a degree. Often used to describe the apparent size of a planet's disk. One arcsecond equals the width of a golf ball viewed from a distance of 8.3 kilometers (5.2 miles).

Astrology. The idea that the positions of the Sun, Moon, and planets determine an individual's personality traits at birth and affect events on Earth. Astrology is largely discredited by the scientific community as a superstitious belief.

Astronomical unit (AU). A measure of distances in space, usually in the solar system. One AU is the mean distance of the Earth to the Sun, and equals 150 million kilometers (93 million miles).

Astronomy. The science that pertains to the study of the universe, its history and future, and the characteristics of the various bodies it comprises.

Aurora. Light emitted by the gases in a planet's upper atmosphere when bombarded by charged particles from the solar wind.

Belts. Atmospheric bands of low altitude and pressure running parallel to a Jovian planet's equator.

Brown dwarf. A body whose mass lies between that of the largest planet and smallest star. A "failed star," it lacks the internal heat and pressure to generate energy through hydrogen to helium fusion.

Catastrophic hypothesis. A theory, now generally discarded, that the planets in the solar system formed when a cosmic body collided with the Sun. The material ejected cooled to form the planets and other orbiting bodies.

Charge-coupled device (CCD). An imaging device that uses an array of light-sensitive elements on a silicon chip. Far more light sensitive than photographic film.

Conjunction. The position of a superior planet when its orbit carries it to the far side of the Sun. Conjunctions of inferior planets are called "superior" when on the Sun's far side and "inferior" when the planet drifts between Earth and the Sun.

Copernican model. A model of the solar system proposed by the Polish astronomer Nicholas Copernicus that the Sun, not the Earth, is located at the center. Also called the heliocentric system.

Core-accretion model. A version of the nebular hypothesis in which all of the planets formed by gathering (accreting) matter from the nebula. In the core-accretion model, the Jovian planets began as rocky cores that accumulated large amounts of gases in the outer part of the nebula.

Deferent. The large circular orbit that a planet makes around the Earth, according to the Ptolemaic model of the solar system.

Density. A measure of how much matter is contained per unit volume, usually expressed in grams per cubic centimeter (g/cm^3).

Differentiation. The separation of materials of varying density in a planet or moon while in a molten state. Denser materials like metals and silicates would settle in the core, while lighter gaseous substances would remain on or near the surface.

Direct motion (orbit). The general counterclockwise motion, both in rotation and revolution, of bodies in our solar system as if viewed from a vantage point high above the Earth's north pole. Most of the bodies in the solar system exhibit direct motion. Also called prograde motion.

Disk instability model. Also known as the gravity instability model, the disk instability model is a variation of the nebular hypothesis in which the Jovian planets formed rapidly from concentrations of gases in the outer part of the nebula, later collecting the heavy materials that formed their cores.

Doppler effect. A shift in the wavelength of radiation emitted by a source that is moving toward or away from an observer. Wavelengths of visible light shift toward the blue (short wavelengths) for an approaching body, toward the red (longer wavelengths) for a receding body.

Electromagnetic energy. A form of energy like light or radio that travels in waves and has an electric and magnetic component.

Electromagnetic spectrum. An arrangement by wavelength of all the energies that travel on electromagnetic waves.

Epicycle. A circular orbit around the deferent, which, in turn, circles the Earth, according to the Ptolemaic model of the solar system.

Extrasolar planet. A planet orbiting a star other than our Sun. Also called an exoplanet.

Gas giant. A huge planet comprised largely of light gaseous materials, primarily hydrogen and helium, with no solid surface.

Geocentric model. *See* Ptolemaic model.

Gravity instability model. *See* disk instability model.

Greatest (eastern or western) elongation. The position of an inferior planet when it appears farthest from the Sun and is most readily seen from Earth.

Heliocentric model. *See* Copernican model.

Horoscope. An astrological forecast of a person's life or earthly events, based on the positions of the Sun, Moon, and planets at a given moment in time.

Hot Jupiters. Giant exoplanets with masses comparable to or greater than Jupiter's, with orbits that bring them extremely close to their parent stars.

Ice giant. A subgroup of Jovian, or gas giant, planets that are less massive and contain a higher quantity of "ices" like water, methane, and ammonia.

Inclined orbit. The high angle of a moon's orbit, relative to the equatorial plane of its parent planet.

Inferior conjunction. *See* conjunction.

Inferior planet. Any of the planets in our solar system whose orbits lie closer to the Sun than Earth's.

Ionized particle (ion). An atomic or molecular particle that carries an electric charge because it has gained or lost one or more electrons. Also called a charged particle.

Jovian planet. A huge, gaseous planet like Jupiter, comprised mainly of hydrogen and helium, with lesser amounts of light gases.

Lagrange point. A stable location in space relative to one or two bodies in which other smaller bodies can safely orbit.

LaPlace Resonance. *See* orbital resonance.

Lenticulae. Based on a Latin term for freckles. Reddish spots and shallow pits about 10 kilometers (6 miles) across that dot the surface of Jupiter's moon Europa. May be caused by upwellings of warmer material from a subsurface ocean.

Magnetosphere. The region around a planet where the motion of charged particles is controlled by its magnetic field, not the solar wind. The side facing the Sun is usually compressed by the solar wind, while the opposite side (the magnetotail) can extend millions of kilometers into space.

Magnitude. A measure of the brightness of a celestial body. The lower the magnitude number, the brighter the object.

Main sequence star. A star like our Sun that shines by converting hydrogen into helium in its core.

Mass. A measure of the amount of matter something contains.

Microlensing. A phenomenon that occurs when the gravity field of a foreground body like a star acts like a lens to bend and brighten light from a more distant background object.

Multiring basin. A large impact area consisting of a main basin or crater surrounded by a series of concentric "bull's-eye" ridges.

Nebula. A vast cloud of gas and dust in space.

Nebular hypothesis. The theory that the solar system (and other planetary systems) was created from a nebula that contracted gravitationally to form the Sun and the bodies that orbit it.

Oblate spheroid. A slightly flattened sphere. This is the shape of the rapidly rotating Jovian planets.

Occultation. The passage of one cosmic body (usually a solar system object) in front of a star.

Opposition. The position of a superior planet when it's opposite in the night sky to the Sun. An opposition occurs around the time of Earth's closest approach to a superior planet.

Orbital eccentricity. A measure of how far an orbit departs from a perfect circle. For the planets and their satellites, orbital eccentricities can range from "0" for a perfect circle to slightly less than "1" for a highly elliptical orbit.

Orbital resonance. A gravitational interaction that occurs between two orbiting bodies if the period of revolution of one is an exact simple fraction of the period of revolution of the other.

Organic compound. A material that contains carbon and hydrogen and usually other elements such as nitrogen, sulfur, and oxygen. Organic compounds can be found in nature, or they can be synthesized in the laboratory.

Palimpsest. Based on a Latin word describing wax-coated writing tablets used by the Romans that could be smoothed and reused. Astronomically, it refers to an impact crater whose topography has been smoothed by glacier-like flow of the surface, typically like ones found on the icy surfaces of Ganymede and Callisto.

Period of revolution. The time it takes a body to complete a single orbit around a larger body, as a planet around a star or a moon around a planet.

Period of rotation. The time it takes a body to rotate, or spin, once on its axis.

Planet. A nonluminous, spherical body in orbit around a star. By definition of the International Astronomical Union in 2006, a planet must be large enough to possess a spherical shape, and it must have gravitationally removed any other bodies from its orbital area.

Planetesimal. One of the numerous small rocky and icy bodies that swarmed around the Sun during the formation of the solar system. Collisions of the planetesimals created the planets.

Plasma. A superheated gas containing positively charged ions and free-floating negatively charged electrons.

Prograde motion. *See* direct motion.

Protostar. The early stage of star formation where the central mass hasn't yet reached an internal temperature and pressure to begin the fusion of hydrogen into helium.

Ptolemaic model. An Earth-centered model of the solar system, created by the Greek astronomer Claudius Ptolemy. Also called the geocentric system.

Radial velocity method. A method of detecting extrasolar planets. As the planet orbits its parent star, a gravitational interaction between the two causes the star to undergo a periodic wobble. A Doppler shift of the star's spectrum as it moves toward or away from the Earth (a radial velocity shift) betrays the presence of the planet.

Radio telescope. A device similar in appearance to a reflecting telescope, except that the reflective mirror collects radio waves from outer space.

Regio. A large area of a planet or moon that is different in color or albedo than surrounding regions.

Retrograde motion. The clockwise motion, either in rotation or revolution, of bodies in our solar system as seen if we were to view the solar system from a vantage point high above the Earth's north pole. Relatively few bodies in the solar system exhibit retrograde motion.

Ring plane crossing. The occasion when the orbits of Earth and a ringed planet allow an edge-on view of the planet's rings. Although all four Jovian planets undergo ring plane crossings, the term generally refers to Saturn's rings.

Roche limit. The distance from a cosmic body at which its gravity would tear apart a smaller body, for example a moon straying too close to its parent planet. Extremely small bodies like meteoroids aren't affected.

Satellite. Any body that orbits a larger body. Often used to describe man-made objects, it also applies to naturally occurring bodies, like moons.

Solar wind. A fast-moving plasma stream comprised of charged particles (mostly electrons and protons) that emanate from the Sun and stream out of the solar system.

Spectrum. A band of colors from red (long) wavelengths to violet (short) interspersed with dark lines, produced when light from a celestial body is passed through a prism, or spectroscope. The bands are produced by the elements in that body.

Star. A huge, hot, spherical mass of gases that generates its own energy through the fusion of hydrogen into helium in its core.

Sublimation. The change in state of a solid directly to a gas without first becoming a liquid.

Superior conjunction. *See* conjunction.

Superior planet. Any of the planets in our solar system whose orbits around the Sun lie beyond Earth's.

Synchronous orbit. The orbit of a body around another in which its periods of rotation and revolution are the same. Bodies in synchronous motion (like our Moon) keep one side always facing the parent body.

Terrestrial planet. A small, dense planet like Earth, comprised primarily of rocky silicates and metals.

Tidal flexing. The periodic bending and warping of a moon's crust, due to gravitational influences by the parent planet or nearby moons.

Torus. A doughnut-shaped mass of material, such as plasma or dust, surrounding a cosmic body.

Transit method. A means of detecting extrasolar planets whose edge-on orbits allow them to periodically pass in front of their parent stars. Extremely sensitive CCD equipment detects the slight dip in the star's overall brightness, betraying the planet's presence.

Volatile. A substance that exists as a gas at normal temperatures.

Zodiac constellation. Any of the 12 constellations through which the Sun, Moon, and planets appear to wander.

Zones. Atmospheric bands of high altitude and pressure running parallel to a Jovian planet's equator.

Bibliography

Arnett, Bill. "An Overview of the Solar System." Available online at: http://www.nineplanets.org/overview.html, April 15, 2008.

———. "Jupiter." Available online at: http://www.nineplanets.org/jupiter.html, April 10, 2005.

———. "Neptune." Available online at: http://www.nineplanets.org/neptune.html, September 2, 2004.

———. "Saturn." Available online at: http://www.nineplanets.org/saturn.html, May 11, 2005.

———. "Uranus." Available online at: http://www.nineplanets.org/uranus.html, October 21, 2008.

Boss, Alan P. "How Do You Make a Giant Exoplanet?" *Astronomy* (October 2006): 38–43.

Bryan, James. "Stephen J. O'Meara and Ring Spokes Before *Voyager 1*" *Journal of Astronomical History and Heritage* (July 2007): 148–50.

Cain, Fraser, and Pamela Gay. "Jupiter." Available online at: http://www.astronomycast.com/astronomy/episode-56-jupiter, October 1, 2007.

———. "Jupiter's Moons." Available online at: http://www.astronomycast.com/astronomy/episode-57-jupiters-moons, October 8, 2007.

———. "Neptune." Available online at: http://www.astronomycast.com/astronomy/episode-63-neptune, November 19, 2007.

———. "Saturn." Available online at: http://www.astronomycast.com/astronomy/episode-59-saturn, October 22, 2007.

———. "Saturn's Moons." Available online at: http://www.astronomycast.com/astronomy/episode-61-saturns-moons, November 5, 2007.

———. "Uranus." Available online at: http://www.astronomycast.com/astronomy/episode-62-uranus, November 11, 2007.

Harvey, Samantha. "Jupiter." Available online at: http://solarsystem.nasa.gov/planets/profile.cfm?Object=Jupiter, May 7, 2008.

———. "Neptune." Available online at: http://solarsystem.nasa.gov/planets/profile.cfm?Object=Neptune, November 13, 2007.

———. "Saturn." Available online at: http://solarsystem.nasa.gov/planets/profile.cfm?Object=Saturn, September 29, 2008.

———. "Uranus." Available online at: http://solarsystem.nasa.gov/planets/profile.cfm?Object=Uranus, June 25, 2008.

Hey, Nigel. *Solar System*. London: Weidenfeld and Nicolson, 2005.

Hoyle, Fred. *Astronomy.* Garden City, NY: Doubleday, 1962.

"Jupiter's Core Twice as Big as Thought." Available online at: http://www.space.com/scienceastronomy/081125-jupiter-core.html, November 25, 2008.

Lin, Douglas N. C. "The Genesis of Planets." *Scientific American* (May 2008): 50–59.

Littmann, Mark. *Planets Beyond.* Mineola, NY: Dover Publications, Inc., 2004.

McAleer, Neil. *The Cosmic Mind-boggling Book.* New York: Warner Books, 1982.

McGourty, Christine. "Lost Letters' Neptune Revelations." Available online at: http://news.bbc.co.uk/1/hi/sci/tech/2936663.stm, April 10, 2003.

Miller, Ron, and William K. Hartmann. *The Grand Tour: A Traveler's Guide to the Solar System*, rev. ed. New York: Workman Publishing, 2005.

Mullen, Leslie. "Uranus and Neptune—and the Origin of Life on Earth." Available online at: http://www.astrobio.net/news/article220.html, June 2, 2002.

"New Images Yield Clues to Seasons of Uranus." Available online at: www.astronomy.com/asy/default.aspx?c=a&id=7503, October 2008.

Pannekoek, A. *A History of Astronomy.* New York: Interscience Publishers, Inc., 1961.

Plait, Phil. "Astrology." Available online at: http://www.badastronomy.com/bad/misc/astrology.html, April 26, 2005.

Porco, Carolyn. "Carolyn Porco Flies Us to the Moons of Saturn." Available online at: www.ted.com/index.php/talks/carolyn_porco_flies_us_to_saturn.html.

Prinja, Raman. *Wonders of the Planets.* New York: Barnes and Noble, 2006.

Proctor, Richard A. *Other Worlds than Ours.* Reprint ed. Whitefish, MT: Kessinger Publishing, 2007.

Rees, Martin. *Universe.* New York: DK Publishing, Inc., 2005.

Sagan, Carl. *Cosmos.* New York: Random House, 1980.

Sobel, Dava. *The Planets.* New York: Penguin Books, 2006.

Wakefield, Julie. "The First New Planet." *Astronomy* (June 2007): 38–43.

Index

About the Author

GLENN F. CHAPLE has been an avid amateur astronomer since the summer of 1963, when a high school friend showed him Saturn with a small reflecting telescope. His astronomical writings have appeared in *Deep Sky Magazine* and *Odyssey*, and he currently writes the "Observing Basics" column for *Astronomy*. He is co-author, with Terence Dickinson and Victor Costanzo, of the Edmund Scientific *Mag 6 Atlas*, and author of *Exploring with a Telescope*. Besides astronomy, Glenn enjoys distance running and fishing. He lives in north-central Massachusetts with his wife, Regina, and is the proud parent of two grown sons and "grampy" to two future astronomers, Katie and Sam.